Amateur Radio
INSIGHTS

Work the World
with D-Star

by

Andrew Barron, ZL3DW

This book has been published by Radio Society of Great Britain of 3 Abbey Court, Priory Business Park, Bedford MK44 3WH, United Kingdom
www.rsgb.org.uk

First Edition 2023

Amateur Radio Insights is an imprint of the Radio Society of Great Britain

The author has no association with any amateur radio equipment manufacturers or resellers.. He has not received any discounts or incentives and will have paid full retail prices for any equipment featured in this book.

ISBN: 9781 9139 9531 7

Cover design: Kevin Williams, M6CYB
Typography and design: Andrew Barron ZL3DW
Production: Mark Allgar, M1MPA

Printed in Great Britain by 4Edge Ltd. of Hockley, Essex

Any amendments or updates to this book can be found at:
www.rsgb.org/booksextra

A note from the author
Many of the examples in the book are based on my observations using the D-Star equipment that I own. There may be variations between the equipment that I am using and your equipment, and that may mean that some things mentioned in the book don't work or need a minor adjustment. But most of the information is generic and will apply to all D-Star radios, hotspots, and services. Where possible I will try to cover any variations. The rules regarding your local repeaters may be different, so it is advisable to talk with members of your local radio club and the local D-Star community.

Work the world with D-Star

Other books by Andrew Barron

Work the World with DMR

Using GPS in Amateur Radio

Testing 123
Measuring amateur radio performance on a budget

Amsats and Hamsats
Amateur radio and other small satellites

Software Defined Radio
for Amateur Radio operators and Shortwave Listeners

An introduction to HF Software Defined Radio
(out of print)

The Radio Today guide to the Icom IC-705

The Radio Today guide to the Icom IC-7300

The Radio Today guide to the Icom IC-7610

The Radio Today guide to the Icom IC-9700

The Radio Today guide to the Yaesu FTDX10

The Radio Today guide to the Yaesu FTDX101

ACKNOWLEDGEMENTS

Thanks to David Grootendorst PA7LIM who writes excellent software for D-Star and the other digital voice modes as a hobby. He makes his software available free to all amateur radio enthusiasts. And thanks to the D-Star repeater and reflector operators who have created a first-class worldwide network. Also, the team at JARL, which pioneered the development of D-Star which ultimately led to amateur radio operators experimenting with all of the digital voice modes. And finally, many thanks to you for buying my book.

ACRONYMS

The amateur radio world is chock full of commonly used acronyms and TLAs (three-letter abbreviations :-). They can be very confusing and frustrating for newcomers. I have tried to expand out any unfamiliar acronyms and abbreviations the first time that they are used. I have assumed that anyone buying a D-Star radio will be familiar with commonly used radio terms such as repeater, channel, MHz, and kHz. Near the end of the book, I have included a comprehensive glossary, which explains many of the terms used throughout the book. My apologies if I have missed any.

Approach

This book is a practical guide that explains the steps that you need to follow to make your new D-Star radio work through your local repeater or hotspot. The learning curve is nowhere near as steep as learning about DMR (digital mobile radio), but there are a few new terms to discover, including dashboards, reflectors, gateways, hotspots, and Echo. Also, acronyms like AMBE+2, DR, DV, CS, and MMDVM. Operating the radio is a little more difficult than operating an FM radio, but with a D-Star radio and a repeater or a hotspot, you will soon be able to 'work' amateur radio stations all over the world.

The book covers how to link to a reflector and what to say when you are making your first calls. If you are using a hotspot you can link to a reflector using the hotspot's Pi-Star dashboard or using the functions on the radio. Or you can use PC software or a phone app. There is a huge amount of excellent information online about D-Star, in the form of websites, forums, pdf files, and 'how to do it' videos. It is not my intention to replace these valuable resources. I hope to complement the insights that they provide and concentrate as much information as possible into a single document.

MMDVM (multi-mode digital voice modem) 'hotspots' are very popular accessories. I have included information about their uses and configuration. I cover the new duplex hotspots and the more familiar simplex hotspots, including a section on how to assemble a hotspot from a kit, plus a Raspberry Pi, and an SD card. This is followed by step-by-step instructions for configuring the Pi-Star modem.

I have also included some information on the D-Star data structure and some observations on the advantages and disadvantages of digital voice technology over FM, and other digital voice modes such as System Fusion, DMR, and P25.

Many readers will already have bought a D-Star handheld or mobile, but for those that haven't, there is a brief guide to the features that you should look for.

The book includes programming instructions for some Icom D-Star radios. The ID-52A, ID-51A +2, IC-705, and IC-9700. Other Icom radios are similar. The instructions provided should cover most D-Star radios, but there might be some minor variations. This section includes programming via the front panel buttons on the radio and using the Icom CS (configuration software).

Icom and others, particularly David Grootendorst PA7LIM, have released some great Android and Apple iOS phone applications and some Windows programs as well. The D-Star terminal modes and routing functions are not often used, but I have included them anyway.

The Glossary explains the meaning of the many acronyms and abbreviations used throughout the book and the Index is a great way of getting to a specific topic.

Conventions

The following conventions are used throughout the book.

D-Star is a digital voice standard employed mostly by Icom radios. Unlike DMR, it was developed specifically for amateur radio use.

DV is Digital Voice, **DD** is digital Data, and **DR** is Digital Repeater.

FM means the normal analog (non-digital) FM mode

Mobile usually refers to a vehicle-mounted radio, but it can just mean a radio that you use while travelling. In this book, 'a mobile' may mean a vehicle-mounted radio or a handheld radio. 'Operating mobile,' means using a radio while you are moving.

Features and settings on the Icom ID-52A include the ID-52E and should broadly apply to the ID-51 and ID-31 variants, and possibly D-Star mobiles.

Features and settings on the Icom IC-705 and IC-9700 may also apply to other Icom D-Star desktop radios.

Fusion, or System Fusion, or Yaesu System Fusion is the digital voice standard employed by Yaesu radios. It is most often abbreviated to YSF (Yaesu System Fusion).

DMR (digital mobile radio) is an ETSI (European Telecommunications Standards Institute) digital voice standard employed by a wide variety of manufacturers. Motorola and Hytera are the biggest commercial vendors of DMR radios and trunked radio repeater systems. Motorola calls their version MotoTRBO.

I have used a `different font` to indicate menu steps in PC software and data entry using the radio buttons.

I use the › symbol to show steps in a menu structure. For example,

Press `Quick › Memory Mode › Group Select`.

I use the ↳ and → symbols to show levels in a menu structure. For example,

`Quick`

- ↳ `Memory Mode → Group Select`
- ↳ `DR Function → GPS Position`
- ↳ `VFO Mode`

'**Click**' or '**left click**' means to click the left mouse button. '**right click**' means to click the right mouse button. 'Click on,' means to hover the mouse over a button or menu option on the PC software and then click the left mouse button to make the selection.

What is D-Star?

D-Star (Digital Smart Technology for Amateur Radio) is a digital voice and data protocol specifically designed for amateur radio. It was developed in the late 1990s by the Japan Amateur Radio League (JARL). Icom was an early adopter, but they don't own the D-Star format. In common with the other digital voice standards, the audio signal from your microphone is converted to a digital data stream with a CODEC (coder/decoder). For D-Star, the digital data from the CODEC, plus other data such as your location and callsign or even pictures, is modulated using GMSK (Gaussian Minimum Shift Keying), which is a variant of FSK (frequency shift keying). The mode transmits a 6 kHz wide signal, taking up half the bandwidth that would be required for a standard FM channel. It was developed as a way of reducing the bandwidth of the transmitted signal while improving the quality of the received voice transmission. Also, the digital voice signal is less prone to noise and flutter fading, resulting in excellent audio quality. This does come at the expense of needing a slightly higher received signal strength than an FM signal. Typically, the signal from a D-Star repeater will sound "perfect" or it won't be received at all. Packet loss or delay in the Internet traffic can sometimes break up or distort the audio from Internet-linked services such as D-Star reflectors, but this is a failure of the Internet routing, not the D-Star format.

D-Star is one of the three most popular digital voice modes alongside DMR and Yaesu System Fusion. There are several other digital voice mode standards used in amateur radio, including P25, NXDN, and Codec2. D-Star is generally considered to be the 'Icom' digital voice mode. This is largely true although other manufacturers notably Kenwood do supply, or have supplied in the past, excellent D-Star radios. Having three popular digital voice modes leads inevitably to questions like "which is best," and "what should I buy?"

WHAT SHOULD I BUY?

The best advice I can offer is to buy the digital voice format that is most popular in your area. If you have a local D-Star repeater, go for D-Star. If most of your friends are using DMR, buy one of those. The same is true for Yaesu System Fusion, P25 or NXDN. You can use any of the digital voice modes through a hotspot. There are many cross-connects between YSF rooms, DMR talk groups, and D-Star reflectors, so you can communicate with amateur operators who are using one of the other modes.

It is generally accepted that DMR is the cheapest. But it is much more difficult to set up and more difficult to use. DMR has the steepest learning curve, but it is also the most popular and it has the most talk groups. You need good computer skills to set up a 'code plug.' Adding another channel using the front panel keypad, for instance, if you visit another town and want to use the local DMR repeater, is quite difficult.

You have to add a new channel for every talk group you wish to use on every repeater. This is not quite as difficult if you have a computer and the programming software handy, but still much more involved than doing the same job on a D-Star or YSF radio.

D-Star is more expensive than DMR because unless you use 'Peanut' or a DV dongle, you are pretty much tied to using Icom radios. With most repeaters and hotspots, any D-Star radio can access any D-Star reflector. D-Star is fairly easy to operate although there are some rules about how to access and use the reflectors.

Yaesu System Fusion is easy to configure. Simply access your local repeater and press the Wires-X button. The radio will list the available rooms and nodes. It is more expensive than DMR because you have to use Yaesu radios. With the latest radios, you can access Wires-X through a repeater or your local network. You can also access some Wires-X rooms and other cross-connected multi-mode talk groups through a hotspot. If you want to host a Wires-X room yourself, you need two Yaesu radios and an HRI-200 box which is an expensive proposition. Yaesu repeaters support C4M digital voice (YSF) and FM at the same time.

WHICH IS BEST?

Once you have made a purchase and set up your radio, I don't think it matters which digital voice technology you chose. There is no clear winner. The three most popular digital voice modes are easy to operate and have similar quality and performance. Many users will tell you that DMR is better because you can use cheap radios, that they prefer D-Star, or that YSF is the easiest to use. It is all very subjective. Some enthusiasts end up with radios for two or all three systems. Several D-Star reflectors, Yaesu 'rooms,' and DMR talk groups are linked together, so no matter which system you choose, you can very easily talk to stations that are using a different DV mode.

The big differences are the price you pay for the radio equipment and the complexity of setting up the radios. The equipment required for each of the digital voice modes is completely different. Each option has 'pros' and 'cons.'

Most D-Star users use the mode because they purchased an Icom radio that includes D-Star. Recent and currently available Icom radios with D-Star include the IC-705, IC-905, IC-9700, IC-7100, ID-5100A, ID-4100A, ID-31, ID-51, and ID-52. Sadly the Kenwood TH-D74A is no longer available. The D-Star mode supports transmitting data files and pictures as well as digital voice. A message which includes the operator's name and often their radio type or city is sent with each transmission. Most D-Star radios include a GPS receiver or the ability to connect an external GPS receiver (IC-9700 and IC-7100). Messages from those radios include the station's GPS location, and the radio can display the distance and bearing to the station you are hearing. On most repeaters or if you are using a hotspot, you will be able to access any D-Star reflector worldwide without having to have them pre-programmed into your radio.

If you want to talk over a linked repeater or reflector, you must register your callsign with the D-Star organisation.

DMR has the advantage that it is a commercial digital voice standard, which means that there are many radios available, ranging from expensive commercial models from Hytera and Motorola to cheaper models from Radioddity, TYT, AnyTone and many other suppliers, mostly from China. The downside is that commercial radios are designed to be set up by a radio network operator or a supplier. They are difficult to program, requiring a channel for every talk group on every RF channel. Once programmed they work just as well as the other radios. You can use a local DMR or multi-digital-mode repeater to access a selection of talk groups. With a hotspot and your radio, you can join a local DMR network or several worldwide DMR networks to access a huge number of talk groups. Some talk groups are international, some support different languages or regions, and some support special interest groups such as model engineering, astronomy, or just chatting over a beer. One thing I like a lot about the Brandmeister DMR network is the centralised dashboard covering every repeater in every location. It shows network activity, repeater information and a lot of other information. You can even listen to active talk groups. The TGIF network has a slightly less capable dashboard with similar functions. Unfortunately having three major DMR networks and a host of regional networks, splits worldwide DMR activity into different camps. I like to use the Brandmeister network, but the New Zealand weekly net is on the DMR+ network. Being a commercial rather than an amateur radio system, your callsign is not transmitted with each call. To display a name and callsign, you have to download a huge user database into the radio to match the stored callsign and name with the calling station's DMR number. Most radios cannot hold enough callsign information for the currently 206,000 DMR users and the database grows bigger as more users are registered. You cannot use amateur radio DMR without registering for a DMR ID number.

TIP: Although all DMR radios support analog FM it is not very good on cheap radios. They usually have a very wide front end, leaving them susceptible to interference. The FM audio quality and levels, both transmitting and receiving, is often very low and "tinny." Don't buy a DMR radio if you want to use it primarily for FM. The Icom, Kenwood, and Yaesu radios used for D-Star and YSF are much better quality, and they work fine on FM.

Yaesu System Fusion (YSF) is generally considered to be the most expensive of the three modes, but the easiest to operate. The expense factor has been alleviated to some extent by the release of the Yaesu FT-70DR handheld which retails at a very reasonable US $175. The Yaesu radios are top quality, but the Yaesu programming software is exceptionally poor. It is extremely difficult to get working and very fiddly to use, especially with the FT-70DR. I was very disappointed with it. YSF is a Yaesu standard. You must use Yaesu radios. One excellent feature is that the Yaesu repeaters support both FM and C4M digital. If a call comes in on FM it is transmitted on FM and the YSF radio will automatically switch to FM to receive it.

If a digital call comes in, it is repeated as a digital signal. This means that FM and C4M digital stations can use the same repeater and C4M (YSF) stations can switch seamlessly between digital and FM conversations.

MYTHS

I often read or hear claims that one digital voice mode has better voice quality than the others. I don't believe that there is any significant difference. Unless you are talking through a repeater directly to another station, I would be surprised if you can tell them apart. As soon as any digital voice mode is passed through the Internet to a remote repeater or reflector, or transcoded to a different digital voice mode, the compression and packet loss in the Internet path will contribute more distortion to the signal than any perceived difference between the different AMBE Vocoder standards being used.

D-STAR COVERAGE COMPARED TO FM ANALOG COVERAGE

Like all radio transmissions on the 2m or 70cm band, you can expect 'line of sight' communications when you are operating outside. If you can see the hill or building the repeater is on, you should be able to use the repeater. The situation is different when you are inside a building, in a built-up area, or in woodland etc. Again, you can expect the coverage of a D-Star repeater to be about the same as an FM repeater.

On the fringes and when mobile, D-Star generally has superior coverage performance to analogue FM because the forward error correction used in the AMBE+2 digital voice CODEC can cope with bit error rates as high as 5% with no degradation in the perceived speech quality. Digital quality is usually better than FM when operating mobile, especially when the received signal is fairly weak. When you are driving, FM often suffers from 'flutter' caused by multi-path propagation. While the received signal needs to be higher for digital reception, the received speech quality should be much better.

BUYING A D-STAR RADIO

Here are some things to consider.

- Desktop, handheld, or mobile. What suits your operating style and budget?
- Dual band UHF/VHF. I believe all Icom D-Star radios are at least dual-band.
- A colour screen that is easy to read and provides plenty of information.
- Waterproof, IPx7 rating or similar if you are planning on hiking or mountain climbing. IPx7 means the radio is protected against temporary immersion in water for 1 m (3.3 ft) for 30 minutes. The Icom ID-52A and ID-51A meet this specification. The IC-705 is not rated.
- Built-in GPS receiver. ID-52, ID-51, and IC-705.

- Bluetooth can be handy if you are operating mobile or portable. You can connect a headset and keep both hands free for driving or climbing.
- A spare battery, or the ability to connect an external battery, can be very important if you are hiking or activating a SOTA peak.

Handheld radio comparison			
Function	ID-31A/E	ID-51A/E plus 2	ID-52A/E
Display size	dot matrix LCD	1.7 inch dot matrix LCD	2.3 inch
Display resolution	unknown	128 x 104	320 x 280 pixels
Display type	Black	Black	Colour
Touchscreen	-	-	-
Menu	Text	Text	Text and icons
Display theme	Fixed	Fixed	Dark/Light
Dual watch	-	VV, UU, VU dual watch	VV, UU, VU dual watch
Dual watch DV-DV	-	-	Monitor a repeater and your hotspot at the same time
Dual watch and FM broadcast	-	Yes	Yes
Bluetooth	-	-	Headset, or RS-MS1 and ST-4100A
Waterfall display	-	-	Yes
D-Star DR function	Yes	Yes	Yes
Picture sharing mode	-	-	Yes
Air band receiver	-	VHF	UHF and VHF
FM Broadcast receiver	-	Yes	Yes
AM Broadcast band	-	Yes	Yes
GPS receiver	GPS	GPS	GPS & GLONASS
Voice recorder	Yes	Yes	Yes
Battery	BP-271	BP-271	BP-271
USB charging	-	-	Yes
Near repeater function	Yes	Yes	Yes
Waterproof	IPX7	IPX7 (IPX4 with BP-273)	IPX7 (IPX4 with BP-273)

Function	ID-31A/E	ID-51A/E plus 2	ID-52A/E
Audio level	400 mW	400 mW	750 mW (louder)
Micro SD card slot	Yes	Yes	Yes
QSO log export	-	.csv file	.csv file
Memory channels	500	500	1000
Scan Edges	50	50	25
Call Channels	2	4	4
DV repeater memory	700	750	2500
FM BC channels	-	500	500
Amateur Bands (transmit)	70 cm	2m 70cm	2m 70cm
Dimensions metric mm	58 w x 95 h x 25 d	58 w x 105.4 h x 26.4 d	61.1 w x 121.6 h x 29.7 d
Dimensions imperial inches	2.26 w x 3.74 h x 1.0 d	2.3 w x 4.1 h x 1.0 d	2.4 w x 4.8 h x 1.2 d
Weight	220 grams	255 grams / 9 oz	330 grams / 11.6 oz
Receiver	double conversion superhet	double conversion superhet	double conversion superhet
Transmit power levels	5, 2.5, 1, 0.5, 0.1 Watts	5, 2.5, 1, 0.5, 0.1 Watts	5, 2.5, 1, 0.5, 0.1 Watts
Frequency stability	±2.5 ppm	±2.5 ppm	±2.5 ppm
Current drain on receive, DV	<450 mA	<450 mA	<450 mA
Current drain on receive, FM	< 350 mA	< 350 mA	< 350 mA

IS THE ID-52 A/E WORTH THE MONEY?

The ID-52 is probably the most expensive amateur band handheld on the market. It is also currently the only D-Star handheld that is available as a new radio. Some say that it is a little too big and that the ID-51 fits your hand better, but for me, I think that the additional features outweigh that concern. The radio has more audio output and a larger speaker addressing one of the main complaints about the ID-51 and ID-31. The colour display has a higher resolution and is much nicer than the old display. Bluetooth enables the use of a cordless headset, and you can use the RS-MS1A, RS-MS1I, and ST-4100A phone apps without a data cable. The colour waterfall is a first for any handheld radio. It could be useful to watch for repeater activity if you have travelled to a new area or are at the top of a mountain. Another first is the ability to listen to two D-Star signals on the same or different bands at the same time.

You can use this to monitor your hotspot and a local D-Star repeater. Finally, it has a picture-sharing mode. You can take a picture on your phone, add a caption, and use ST-4100A to copy it to the radio over the Bluetooth connection and send it via D-Star.

Desktop radio comparison			
Function	IC-9700	IC-705	IC-905
Display size	4.3″ colour TFT	4.3″ colour TFT	4.3″ colour TFT
Display type	Colour	Colour	Colour
	IC-9700	IC-705	IC-905
Display size	4.3-inch colour TFT	4.3-inch colour TFT	4.3-inch colour TFT
Display type	Colour	Colour	Colour
Touchscreen	Yes	Yes	Yes
Display resolution (pixels)	480x272	480x272	480x272
Menu	Text and icons	Text and icons	Text and icons
Display theme	Fixed	Fixed	Fixed
Dualwatch	any 2 bands	-	-
Dualwatch DV DV	-	-	-
Bluetooth	-	Yes	Yes
Waterfall display	Yes	Yes	Yes
D-Star DR function	Yes	Yes	Yes
D-Star DD function	Yes	No	Yes
Picture sharing mode	Yes	Yes	Yes
ATV (FM)	-	-	Yes
Satellite mode	Yes	-	?
Marine band receive	-	Yes	?
Air band receive	-	Yes	?
FM BC receive	-	Yes	?
AM BC receive	-	Yes	?
GPS receiver	External	Internal GPS	Internal GPS
Voice recorder	Yes 8 channels	Yes 8 channels	Yes 8 channels
Battery	-	BP-272	BP-272
USB charging	-	Yes	Yes
Waterproof	No	No	No (ODU yes)
Audio level	2W	530 mW	530 mW
Micro SD card slot	Yes	Yes	Yes

Function	IC-9700	IC-705	IC-905
QSO log export	.csv file	.csv file	.csv file
Memory channels	99 per band	500	?
Scan Edges	6 per band	25	?
Call Channels	2 per band	4	?
DV repeater memory	2500	2500	2500
Amateur Bands (transmit)	2m, 70cm, 23cm	HF, 2m, 70 cm	2m, 70cm, 23cm, 12cm, 5cm, 3cm (opt)
Modes	SSB, CW, RTTY, AM, FM, DV, DD	SSB, CW, RTTY, AM, FM, DV, DD	SSB, CW, RTTY, AM, FM, DV, DD, ATV
Dimensions metric mm	240 w x 94 h x 248 d	200 w x 83.5 h x 82 d	200 w x 83.5 h x 82 d
Dimensions imperial inches	9.4 w x 3.7 h x 9.4 d	7.9 w x 3.3 h x 3.2 d	7.9 w x 3.3 h x 3.2 d
Weight	4.7 kg / 10.4 lb	1.1 kg / 2.4 lb incl battery	?
Receiver	SDR, Hybrid SDR on 23cm	SDR	SDR, Hybrid SDR on 23cm and above
Transmit power levels	100W on 2m, 75W on 70cm, 10W on 23cm	10W	10W on 144 / 430 / 1200 MHz; 2W on 2400 / 5600 MHz; 0.5W on 10 GHz transverter
Frequency stability	±0.5 ppm	±0.5 ppm	GPS locked
Current drain on receive, DV	1.2A	0.5 A standby 0.8A RX	unknown
Current drain on receive, FM	1.2A	0.5 A standby 0.8A RX	unknown

THE IC-905 PROMISE

The IC-905 has not been released yet (Oct 22). It is another great example of Icom pushing the limits of amateur radio and offering something completely new and exciting. The radio extends D-Star capabilities onto the bands above 23cm, but I doubt there will be much activity. I am wondering if the radio will have full duplex cross-band satellite capability for users of the QO-100 geostationary satellite. Other possibilities are microwave EME and long-distance challenges. I expect that the radio will be expensive given the capabilities that it offers. But I still want one!

How to talk on D-Star

The next few chapters include all the technical information about setting up your D-Star radio and about reflectors and gateways and other cool stuff. But I thought I would take some time to discuss what you should say and how you go about making a 'contact' with another station. This can be daunting for new Hams and chatting with an amateur radio operator from another country may be a new experience for some readers. Dive right in, ham radio people are friendly!

You do need to be aware of cultural and political differences, different accents, and the fact that English may not be the first language of some of the folks out there. Think about how you would feel if you moved to Europe or Asia. I am willing to bet their English is way better than your Korean, Japanese, or Portuguese. But we all have one thing in common and that is a love for amateur radio. It is important to remember when you are talking on a reflector, that you are not just talking to people from your neighbourhood. There could be hundreds of people in dozens of countries linked to the same reflector. This is especially important if you use picture-sharing. Be aware that anyone on the reflector that has a picture-capable radio will automatically receive any image that you send, whether they want it or not. There are no private conversations on D-Star.

Preliminaries

You need to register for D-Star and add your hotspot frequency or local repeater frequency to the DV memory in your transceiver. Your local repeater may already be listed in the DV memory. If it is, just select it. The setup process is fully covered later.

The next few steps are in the 'Fast track – minimum configuration' flow chart, in the next section. They are, select a reflector, link to the reflector, change to the 'Use Reflector CQCQCQ' mode, and talk.

Listen first

It is a good idea to listen to one of the busy reflectors to get an idea of how the calls are made and answered. Try REF001 C or REF030 C.

Making your first D-Star call

The easiest thing to do for your first call is to "come back" to another station.

For example, if you hear F2YT say they are "listening" or "listening for any calls," feel free to call them using a format like, *"F2YT this is ZL3ADB"* or *"F2YT this is ZL3ADB my name is Andrew, and I am located in Christchurch."*

Stations that key up

Every station that transmits onto the reflector, will show up on the screen of your radio, your hotspot display and dashboard, or the repeater dashboard. You usually get a voice announcement of their callsign. They may not say anything.

On your radio, you will see their callsign followed by the operator's name and location and often the GPS position screen. You may need to open that by touching the red (or grey) compass icon. The stations may not be looking for a call. They might be testing, letting a friend know they are on the reflector, or shy. Feel free to give them a call. Most stations will respond. If they are anything like me, it might take a minute for them to get organised and they probably won't have your callsign correct yet. If they don't wish to have a contact with you, they won't answer. Don't be surprised or disappointed if someone does not respond to your call. They might be busy with something else. You will probably find that you do the same thing.

The format for calling them is the same as above. Take note of their callsign and give it a try. *"W2ABC this is 2E0AB"* or *"W2ABC this is 2E0AB my name is Bob, and I am located in Christchurch England."*

TIP: Some stations transmit a G-PRS beacon to the aprs.fi website every ten minutes or so. The operator may not even be in the room, or they might be busy hiking or driving.

Initiating a call

If nobody is talking, you can certainly initiate a call. Most people don't call "CQ" on repeaters and reflectors, but there is no rule that says you can't. I believe that it is a good idea as it makes it very clear that you are looking for a contact from anyone, anywhere. Don't make a long CQ call. There are no band conditions or noise to worry about, and people are not tuning across a band looking for signals.

There is no need to use the phonetic alphabet because there is no noise to mask your signals. But some people like to use it anyway. I only use phonetics if a station is having a hard time getting my callsign right.

A good format would be, *"JA5GG listening on 001 C,"* or *"JA5GG listening for any calls,"* or even *"JA5GG listening for anyone anywhere."*

You can call CQ. *"KE6DQ calling CQ on 1 Charlie"* is fine. I like to mention my callsign at the end of the sentence because a station will hear the CQ and then listen for the callsign. Try *"CQ on 1 Charlie, this is KE6DQ."*

If you are interested in talking to someone in a particular country or about a particular topic, you can mention that. *"G3ES looking for Florida,"* or *"EA8TL looking for help with sending pictures,"* or *"Anybody interested in ballroom dancing? This is DL3AA."*

Stations that don't key up

There may be hundreds of people just listening to the chatter or waiting for a friend to show up. Unless you are watching a reflector or gateway dashboard you probably won't know how many are 'out there.'

FAST TRACK – MINIMUM CONFIGURATION

If you are familiar with adding channels to your radio and you are "ready to go," here is the fast way to get "on the air" with D-Star. The key point is to change to 'Use Reflector CQCQCQ' after you have established a link to the reflector or gateway.

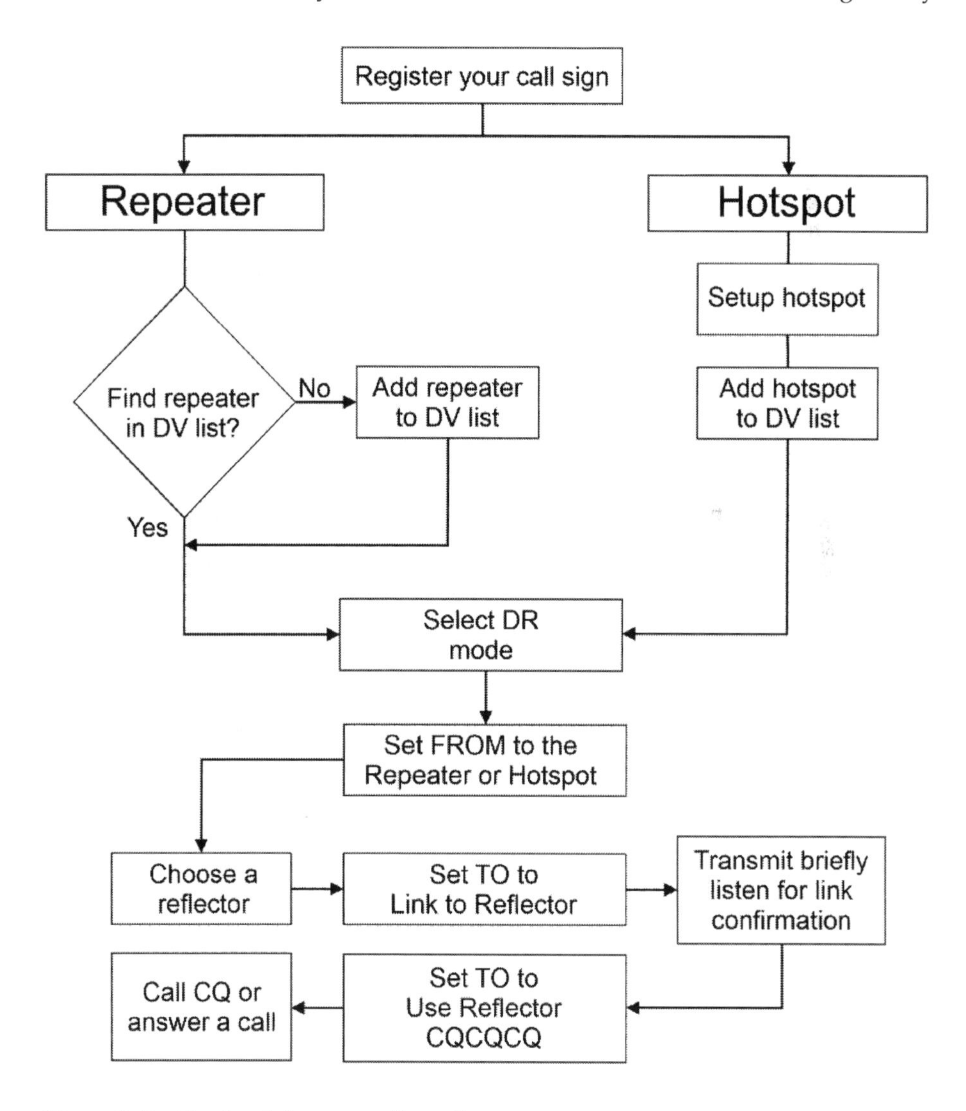

Figure 1: Fast track minimum configuration

Reflectors and modules

REFLECTORS

A reflector is the D-Star equivalent of a DMR 'talk group' or a YSF 'room.' Of the three terms, I think that 'room' best describes the function. You can also think of it as a repeater that is hosted on the Internet or a meeting place where like-minded people hang out. Reflectors extend D-Star operation outside the coverage of your local repeater, allowing you to talk with amateur radio operators all over the world. They are often used for special interest group 'nets' because you can reach so many people.

You can use commands on the radio or a phone app to establish an Internet link between your local repeater and a 'reflector.' If you are using a hotspot, you can use the radio controls, a phone app, or the hotspot dashboard.

Reflectors often have interconnections with other D-Star reflectors and with other digital voice modes. See the picture on page 17 showing the many interconnections associated with the XLX299 reflector. You cannot choose which other repeaters are connected to a reflector, but you can choose the reflector you want to connect to. Users within each repeater coverage area can connect or disconnect their repeater or hotspot as they wish. It is good etiquette to ask before you disconnect your local repeater from a reflector. Someone might be waiting for a call or for a net to begin. You can connect or disconnect your hotspot as often as you like.

Once the link has been established you can put out a CQ call, respond to others using the distant repeaters, or call a station that is within the coverage of one of the connected repeaters. Any call you make to another station is not private. It will usually be broadcast on your local repeater and all the other connected repeaters and hotspots worldwide. If you established the connection to a reflector, you should probably reinstate any link that was in effect before you started. If you didn't establish the connection and you don't want to make a different connection, you should leave the link connected. The rules may be different for your local repeaters. Your radio does not have memory slots for reflectors, but it does store the last few that you have used. You can also create a list of favourites in the 'Your Call Sign' table.

GATEWAYS

A gateway node is a repeater or hotspot that is connected to the Internet. The gateway allows you or the repeater owner to connect your local repeater to a reflector across the world. You can also create a direct link via the Internet between your hotspot or local repeater and another D-Star repeater. Once the link is established you can put out a CQ call, respond to others using the distant repeater, or call a station that is within the coverage of the distant repeater. If the far repeater is linked to a reflector, you will hear the reflector traffic as well.

But there is a 'snag.' The **Gateway CQ** mode and **Direct Input (RPT)** modes on your radio will not work on most hotspots and non-Icom repeaters. These methods use routing rather than linking and this is not supported on hotspots. However, you can establish a link (not a route) to a particular gateway repeater. See page 49.

Most D-Star users never use gateway routing or linking. They stick to monitoring and calling on the reflectors. Gateway links are most useful when you know the repeater that your friend or contact is using, and you want to chat with that person in particular. Even then many people simply listen at a particular time on an agreed reflector.

Since gateways are repeaters, they have an RF component. They can transmit and receive RF transmissions. Reflectors do not have an RF component. They exist only as software running on a server.

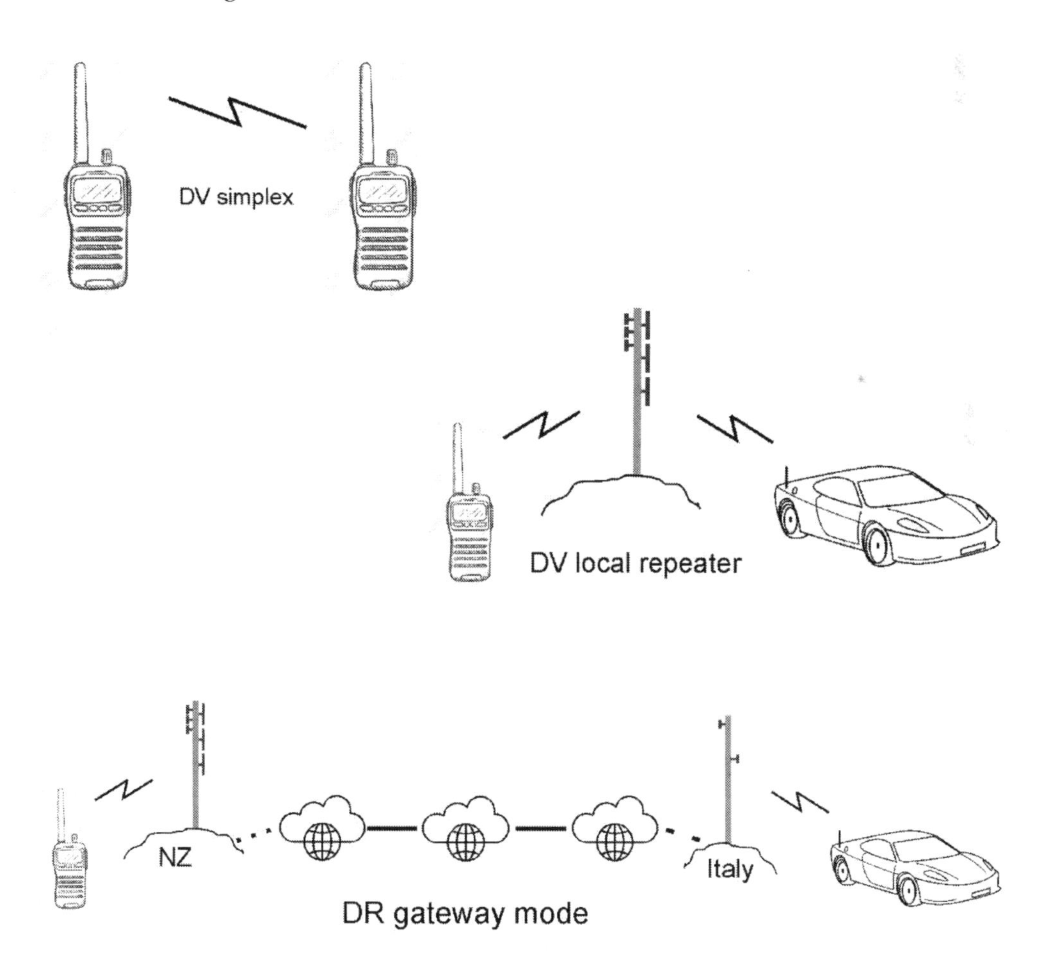

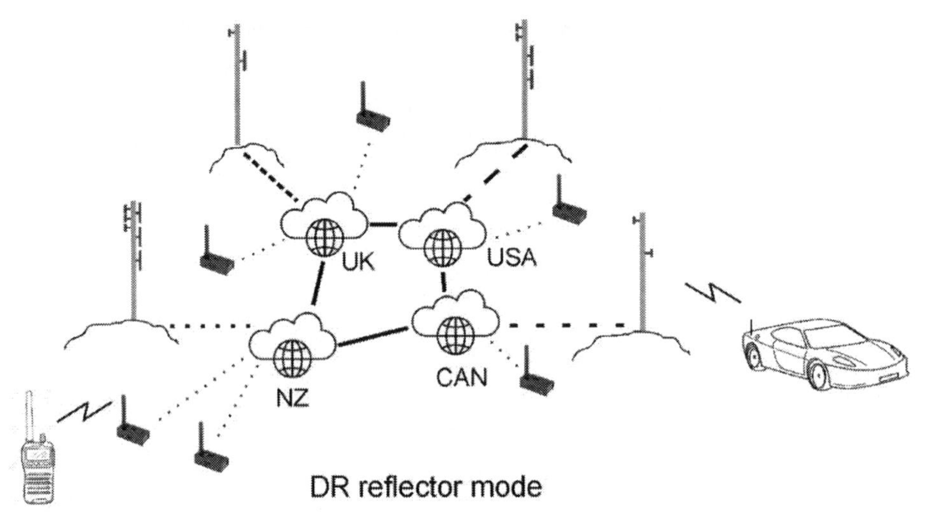

Figure 2: D-Star operating modes

MODULES

You may have noticed that many D-Star callsigns include a suffix letter. It is called a module and it is always in the 8th position of the text field. Your hotspot has one. It is usually a B indicating that it is working on the 70cm band or a C indicating that it is working on the 2m band. You can use any letter that is not reserved, but there are preferred codes. If you register two D-Star radios against your callsign, one of them can use your callsign without a suffix. The other must use a module code so that the system can tell them apart. I am using H for my handheld. If you have two hotspots, they must have different module codes. You can in write yours or add any I missed.

Suffix	Definition	Suffix	Definition
A	23cm band	N	
B	70cm band	O	
C	2m band	P	
D	Data traffic/modem	Q	
E	Echo (reserved)	R	
F		S	System (reserved)
G	Gateway (reserved)	T	Terminal mode
H		U	Unlink (reserved)
I	Information (reserved)	V	
J		W	
K		X	
L	Link (reserved)	Y	
M		Z	

Reflector modules

Reflectors have a callsign so that you can link to them. They can host as many as 26 modules, using use all the letters from A-Z. Since a reflector does not transmit, there are no constraints on what the letters stand for. However, traditionally A is used for an international talk group, B for a national talk group, C-F for regional talk groups, and higher letters for other special interest groups.

Reflector modules are sometimes called ports. You might hear someone say, "I called Fred on REF001 port C."

Traditional REF reflectors can have up to five modules with E reserved for Echo. XLX, DCS, and XRF reflectors can have up to 26 modules. I and U are usually used for Information and Unlink. Module E might be used for Echo. Some module letters might be reserved for permanent links that mere mortals like us cannot connect to.

This diagram shows just how complicated reflector linking can get. It is a picture of the XLX299 reflector which has ten attached modules. There are D-Star, YSF, AllStar, DMR Brandmeister, DMR+, TGIF, ZL MOTOTRBO, and FreeStar connections. For example, you can connect to Module B from a D-Star repeater or hotspot, or via DMR BM TG530, module C on XLX530, YSF room 62078, or a Peanut server.

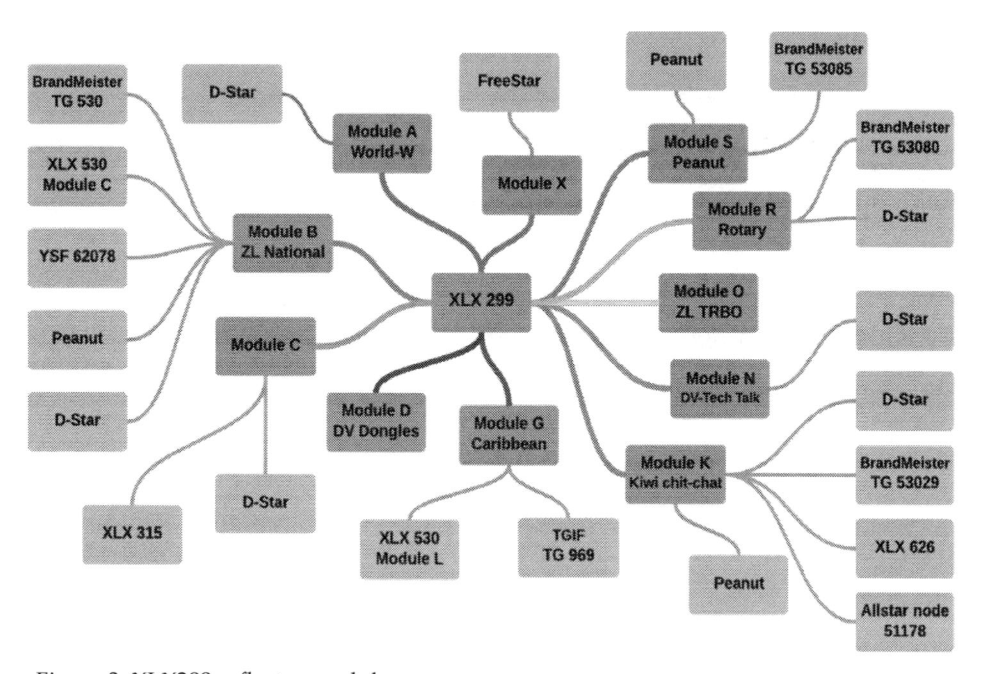

Figure 3: XLX299 reflector modules

Image from: https://www.nzart.org.nz/activities/dv-technical-talk-net

Repeater stacks

The modules were originally designed for situations where a single D-Star repeater controller was controlling repeaters on several bands. Each band has a module code and the whole station is called a repeater stack. Some repeaters have several modules but only one or two RF frequencies. In that case, the other modules can act as a sort of mini reflector. Two or more stations can connect to the module and talk among themselves without the repeater transmitting the signal over the coverage area.

D-STAR REFLECTOR TYPES

Older radios support linking to D-Star reflectors that have REF numbers. Newer radios can also connect to multi-mode reflectors using XRF, DCS, or XLX numbers. Some reflectors use multiple codes. For example, connecting to DCS 299 is the same as connecting to XLX 299. Some reflectors have permanent connections to modules for other digital modes. This allows you to talk to DMR, P25, or System Fusion users.

TIP: There is a workaround that lets all D-Star radios access any of the reflector types.

REF

REF is the DPLUS reflector system. It is a closed-source proprietary system developed by Robin Cutshaw AA4RC. REF reflectors were the first generation of D-STAR reflectors and the standard is still very much in use. There is a list of REF reflectors at http://www.dstarinfo.com/Reflectors.aspx.

- REF001C in London is D-STAR's "Mega Reflector." It usually has many repeaters connected to it.
- REF030C in Atlanta is also very popular.

XRF

XRF is the Dextra X-Reflector system, created by Scott Lawson, KI4KLF. It is the second generation of D-STAR reflectors, and it is open source. A list of XRF reflectors is available at http://www.xrefl.net/.

DCS

DCS is the 'Digital Call Server' reflector system. It was developed by Torsten Schultze, DG1HT, and is now run by Stefan, DL1BH, Peter, DG9FFM, and Rolf, HB9SDB. See http://www.xreflector.net/

XLX

XLX is the newest reflector system. It was developed by Jean-Luc Boevange, LX3JL, and Luc Engelmann, LX1IQ. XLX is an open-source system, which supports all of the other D-STAR protocols like DCS, XRF (Dextra), and REF (DPlus), as well as other digital modes. The XLX system is often used to connect reflectors and to interface with YSF rooms, DMR talk groups, and other digital voice modes. http://xlx000.xlxreflector.org/index.php?show=reflectors.

REFLECTOR DASHBOARDS

Dashboards are Internet websites associated with gateway repeaters and reflectors. They are a great way of working out when reflectors are busy and checking that you are getting out onto the D-Star network. If you transmit to the reflector, your callsign should show up on the last heard or users list. They also show which modules the node accepts and what they are linked to.

It's worth checking out the first few links below just to see what special interest groups and countries host the reflectors.

There is a list of **XRF reflectors** at http://xrefl.net/. Clicking the blue link should take you to the specific reflector dashboard.

D-Star Info has a list of **REF reflectors** at http://www.dstarinfo.com/reflectors.aspx

Most **XLX reflectors** have a tab showing a list of active XLX reflectors. I am not sure how up-to-date they are. Some sites show reflectors as being active while others show them as inactive. http://www.xlx750.nz/index.php?show=reflectors

Peanut global dashboard http://peanut.pa7lim.nl/

DCS Reflectors http://xreflector.net/ Scroll to the bottom of the DMR and YFS links on the left. Most are multi-protocol.

Reflector	Country	Modules	Reflector	Country	Modules
DCS001	Germany	26	DCS015	Germany 2	7
DCS002	World-Wide	26	DCS017	Brazil	26
DCS003	Switzerland	1	DCS018	Spain	8
DCS004	Denmark	3 (B,C,S)	DCS019	Czech	16
DCS005	United Kingdom	23 (B-Y)	DCS022	Japan	?
DCS006	USA	26	DCS023	Bulgaria	2 (A, B)
DCS007	Netherlands	9	DCS026	USA 2	4 (A-D)
DCS009	Austria	6	DCS032	Luxembourg	2 (A, B)
DCS010	Sweden	5	DCS033	France	3 (C,N,R)
DCS011	Belgium	5	DCS262	Ham Cloud	?
DCS012	Portugal	?	DCS945	Germany 3	22
DCS014	Australia	1 (B)			

REF001 http://ref001.dstargateway.org/ has the usual five modules, but the only one that gets any action is module C, which is the 'Worldwide D-Star mega reflector.' REF001 C is the busiest D-Star reflector in the world. There is a last-heard list at https://dstarusers.org/viewrepeater.php?system=ref001

REF030 http://ref030.dstargateway.org/ is the second busiest D-Star reflector. It is sponsored by Georgia D-Star in the USA. It seems to have activity on module C and module B. There is a 'last heard' table at the bottom of the web page. When I checked, 873 people were listening and nobody was talking.

The 'DARA Thursday Night Group' website has links to 86 D-Star REF reflectors https://sites.google.com/site/darathursdaynite/d-star/d-star-dplus-reflectors

XRF757 A https://xrf757.openquad.net/ is the QuadNet Array. It links QuadNet Smart Groups DSTAR1 and DSTAR2 to D-Star Reflectors, Brandmeister DMR talk groups, DMR plus talk groups, YSF rooms, and Wires-X rooms.

XRF757 C is QuadNet Tech Chat

XLX750 http://www.xlx750.nz/ has 21 modules. You can link to 14 Brandmeister DMR talk groups, 4 DMR plus talk groups, a P25 bridge, XLX049 Ireland, a FreeSTAR DMR talk group, and QuadNet DMR TG320.

XLX299 https://www.xlx299.nz/index.php?show=modules has 10 linked modules. Six are available to peanut users. *BM TG means a Brandmeister DMR talk group.*

- XLX299 A is XRF899 and also UK XLX899
- XLX299 B is BM TG530, Peanut XLX299B, and NZ National XLX350
- XLX299 D is DV D-Rats and SSTV
- XLX299 G is TGIF DMR 969, Peanut TGF969, DMR Caribbean
- XLX299 K is BM TG53029 Kiwi DMR chat channel
- XLX299 N is NZART (New Zealand) DV-Tech talk
- XLX299 R is BM 53080, Peanut DCS299R, ROAR YSF reflector
- XLX299 S is BM 53085, Peanut DCS299S, Active Elements
- XLX299 U is UK D-Star
- XLX299 X is Peanut DCS299X, FreeSTAR Module X

XLX049 https://xlx049.celticcluster.org/ is a multi-protocol reflector.

- XLX049 B is Ireland's regional chat group
- XLX049 D is a link to QuadNet DTSAR1
- XLX049 B is a link to QuadNet Smart Net QDVNI2

Getting started

The first thing you need to do is check out what D-Star repeaters are available in your area. It is a good idea to talk to D-Star users at your local amateur radio club. If there are no D-Star repeaters but you have good a good Internet connection, you can buy or build a hotspot and use that to communicate with radio hams on reflectors and gateways all over the world. Next, you need a D-Star handheld, mobile, or desktop radio. Take a look at some things to look out for, on page 6 in the 'What is D-Star' chapter. Kenwood used to make some D-Star models but now it seems to be only Icom radios. I have concentrated on the IC-705, IC-9700, ID-51A +2, and ID-52A. Programming and using earlier handheld radios and Icom mobile radios such as the ID-4100 and ID-5100A should be similar. The book includes setup instructions and how to make your first few calls. 'E' models such as the ID-51E and ID-52E are for Europe. Transmitting is limited to 144-146 MHz and 430-440 MHz.

Once you have your radio, there is a little work to do before you can use it. You need to enter and store the frequencies of your local repeater or hotspot, as you would for an FM transceiver, and you should register your callsign on the D-Star registry. After that, you are "good to go." It may seem rather daunting at first because of all the new terms you need to learn and having to adjust to a different way of operating the radio. But we will step through the process, and it will soon become second nature.

Unlike DMR where every talk group must be programmed for every RF channel, D-Star radios can access all D-Star reflectors and gateways without you having to download any database files. You don't have to download a callsign database either. The radio will display the callsign and name of anyone you are hearing. A D-Star radio only needs one channel for each repeater or hotspot. You can travel to a different town or country, enter the frequencies of the local D-Star repeater and use your favourite reflectors very easily.

Your radio might already have a populated D-Star repeater list. But my IC-9700 was supplied without any repeaters loaded. You only need your local D-Star repeaters on the list. It is helpful to have a wider list if you travel because the radio can automatically show you repeaters that are close to your location. I do recommend adding your hotspot to the repeater list. It allows you to use all the features of the DR (digital repeater) mode. It is not 100% necessary to use the DR mode, but it is the method I will cover. You can also add some nearby FM repeaters if you wish.

I describe changing settings directly on the radio and using the free Icom CS configuration software. Some functions are easy to change using the radio buttons, others, such as adding a lot of FM repeater channels are easier using the software. I have not covered the RT Systems software this time because it is very similar to the free Icom software. Any data that can be entered into the radio can also be entered using the CS software. Use whichever method you find easiest.

ENTER YOUR CALLSIGN

The very first thing that you must do is enter your callsign into the radio. No long distance digital mode conversations can be made unless this simple step is done.

IC-705, IC-9700 and similar radios

Use **MENU > Set > My Station > My Call Sign**. Touch and hold the top position and then touch **Edit**. Enter your callsign before the / symbol. After the / symbol you can enter a four-digit code. It is usually the radio type. For the IC-9700 you would enter 9700. But it could be P for portable, M for mobile, VK if you are temporarily operating in Australia, or K4 if you are temporarily operating in the K4 call area. Some people put their name after the / symbol. But there is a better place for that.

You can enter up to eight callsigns. Perhaps more than one operator will use the radio, or you may want to add separate entries for /9700, /P, and /M. Most people will only use one 'My Callsign' entry.

Press **ENT** to save the changes or they will be lost. Then exit using the ⮌ Soft Key twice or the **EXIT** button twice to get back to the 'My Station' setup screen.

On the **TX Message (DV)** line you can enter the information that will be displayed when other stations see and hear your transmission. Most people put their name and location (*Andrew Christchurch*) or their name and Maidenhead grid reference (*Andrew RE66hm*). **MENU > Set > My Station > TX Message (DV).**

By the way, after your callsign has been entered. It will be displayed on the Icom splash screen when you turn on the radio.

If you are configuring an IC-9700 you may notice that the 'My Call Sign (DD)' entry has been updated to show the same callsign as you entered for the DV mode. You can change it if you want to.

ID-52, ID-51, and similar radios

The callsign configuration for the ID-52 and ID-51 is basically the same, although it might be easier to use the CS (configuration software) because it is fiddly to enter text on the radio.

Use **MENU > Set > My Station > My Call Sign** then press the **Quick button > 1 > Edit**. Enter your callsign, press the **right (LO) button** until the cursor is after the / symbol and enter a four-character suffix, usually something like ID52 or ID31, and save the data with the **ENTER** key.

TIP: the ***up*** *(RX-CS) button and* ***down*** *(DR) buttons select the character, the* ***right*** *(LO) and* ***left*** *(CD) buttons move the cursor, and* ***VFO/MEM*** *clears a character. The* ***centre button*** *is the ENTER key.*

D-STAR REGISTRATION

You can make local repeater calls and point-to-point simplex calls to another D-Star radio without any further complications. But if you want to talk over a reflector or a linked repeater, you must register your callsign with the D-Star organisation. It is an online process done from your Internet web browser, rather like registering for an Internet forum. Most people register with a repeater near them because it becomes their 'home repeater.' Your home repeater is your default location if the system has no 'last heard' record of the reflector you are using.

Icom has released a video covering the process, but I will step you through it anyway. https://www.youtube.com/watch?v=cPp8DHB9arQ

On your PC, enter dstarinfo.com into the URL line on your Internet web browser.

Click on **Repeater Maps & List > Repeater List**.

Select your **geographical area** using the dropdown list box. If you are in New Zealand, select Oceania. The list shows a handy list of D-Star repeaters and the modules they support. **Info** provides information about the repeater, provided by the repeater owner.

Select a repeater near you that has a '**Registration**' link and click the link. If you cannot find a suitable link, you can register at https://regist.dstargateway.org/ This site also has a link to the registration instructions.

Usually, registration is a two-stage process, but some repeater moderators have automated the process. You enter your callsign and create a password. In a day or two, you should get an email from the repeater moderator. Armed with the required information you go back to the website and enter a small amount of additional information. Log in and click **Personal Information** (some websites skip this step). Check the checkbox on the first line. Enter your callsign in uppercase (it will probably be pre-populated). In the box marked **Initial** enter a space. You **must** enter a single space. Leaving the field blank will not work. Leave **RPT** unchecked. In the **pcname** field enter your callsign a hyphen and a short radio type, all in lowercase. This **must** be in lowercase. For example, **zl3dw-id31a**. Or you can just enter your callsign without the extension. Don't put your name here.

You will notice a 'local IP' address has been generated. You don't need to remember it. You won't need it again unless you are using an ID-1 or a dongle server.

If you operate two D-Star radios, for example, a mobile and a home station, you can fill in two lines, one for each D-Star radio that you are registering.

You would also complete an additional line if you were planning to use the radio as a public gateway or repeater. You can always do it later.

Press the '**Update**' button to complete the process.

Congratulations! You will never have to do this again unless you move permanently to a new location, or you need to register another radio or a hotspot.

Checking your D-Star registration

You can check your D-Star registration at the Pi-star UK website. Enter your call and it will show you where and when you registered for D-Star. https://www.pistar.uk/d-star_regcheck.php.

https://regist.dstargateway.org/regcheck/index.php has the same information.

RADIO MEMORY BANKS

The ID-52 can store 1000 channels in 100 banks. The ID-51 +2 holds 500 channels in 26 banks. The IC-705 can store 500 channels in 100 banks. The IC-9700 has a 99-channel memory bank for each of the three bands that the radio can operate on. These memory slots can hold FM channels and digital channels, but I prefer to use the DV memory bank for digital repeater channels. On the IC-9700 it is the DV/DD memory bank.

REPEATER LIST - DV OR DV/DD MEMORY

The DV memory is primarily for D-Star repeaters. Recent radios can hold 2500 DV repeater channels in 50 groups. Each group usually holds the repeaters for a particular country or region, but you can add a new group to hold your favourites. Some Icom D-Star radios are supplied with a worldwide repeater database already loaded. If you are using a hotspot, you should add it to your country or regional list. The ID-51 +2 can store 750 channels in the DV repeater list.

Routing might work in Japan where almost all the repeaters are made by Icom. But having a large list of D-Star repeaters is pointless for most users because routing to them does not work. You only need your local repeaters, the repeaters in areas you might travel to, and your hotspot in the DV memory. You can add FM repeaters to the list as well. I included my two local FM repeaters so that I can check the activity on them without exiting the DR mode. I can also scan them along with the local DV repeaters. The Icom manual states that the DV or DV/DD memory bank is pre-programmed, but my IC-9700 radio was supplied blank. Check to see if there are records in the DV memory bank using, **Menu > DV Memory > Repeater List >** group.

If your radio already has the worldwide repeater list, you can just add your hotspot if you use one. If your radio does not have the worldwide repeater list, you can download a .csv file containing the D-Star repeaters from the download area of the Icom Japan website, https://www.icomjapan.com/support/firmware_driver/2407/.

Or you can just add the four or five channels you need. Either manually add them using the radio controls or using the Icom CS (configuration software).

You could download the full list, edit it with a spreadsheet program to remove all repeaters that are out of range, and then upload it to the CS program and your radio using either of the following methods.

If you do want to load the complete list

If you do want to load the complete list or an edited list, there are two ways to get the file into your radio. The easiest way is to upload it using the CS program. The less direct way is to copy it onto the radio's SD card and then import it.

*TIP: Not all D-Star radios have an SD card and not all D-Star radios support importing a .csv file. The IC-7100, ID-31 and ID-51 radios have the repeater information embedded in an .icf file. The .icf file is what the CS software saves. Copy the .icf file to the '**Setting**' subdirectory on the SD card and load it from there. The ID-31+ and ID-51+ and newer models do support importing a .csv file.*

Update the repeater list – CS program method (preferred)

Download the file from the link. Edit the file with a spreadsheet if desired. https://www.icomjapan.com/support/firmware_driver/2407/.

Connect the radio to the PC with a USB cable or the OPC-2350LU data cable as appropriate for your radio and start the CS programming software. See 'Configuration software,' starting on page 72.

Click on the **Clone read** icon (arrow pointing to a computer) to read the current radio configuration.

Click **Digital** then right-click the **Repeater List** folder to reach the import function. Select **Import > All**. Navigate to the Windows folder where you saved the .csv file and click **Open**.

TIP: It is possible to update just one group using this method. Right-click the Group Name rather than Repeater List. You can import or export a group or the entire .csv file.

*TIP: Note that the CS-51 software does not support a right mouse click. You have to select the target folder and then use **File > Import > All or Group**.*

Click on the **Clone write** icon (arrow pointing to handheld radio) to write the new radio configuration back to the radio.

Finally click on **Save**, to save a backup on your PC.

Update the repeater list - SD card method

Remove the SD card from the radio and install it in your SD card reader attached to your PC. See page 70 for SD card details.

Download the .csv file to a Windows subdirectory and then copy it onto the *radio model*\CSV\RptList directory of the SD card.

Carefully put the SD card back into your radio then import the file using
MENU > SET > SD Card > Import/Export > Import > Repeater List > D-Star List. Answer the Keep SKIP settings question with YES.

ADD YOUR HOTSPOT TO THE REPEATER LIST

If you are using a hotspot, I strongly recommend adding it to the DV memory repeater list. This can be done directly on the radio or via the CS program. You can create a new group for your hotspot(s), but it is unnecessary.

I do recommend filling in the latitude and longitude data because it will make your hotspot appear in the 'near repeater' list, saving you the bother of looking through the whole list for it.

Add your hotspot to the repeater list using the CS program

Connect the radio to the PC with a USB cable or the OPC-2350LU data cable if appropriate and start the CS programming software. See 'Configuration software,' starting on page 72.

Click the + next to the Digital folder then the + next to the Repeater List. Find the group for your country or region. Or click on Group Name if you want to add a group.

Right-click any cell in the top, or any row, and select Insert. This will add a blank line. Fill in the details. You can copy-n-paste some information from other lines.

Type: DV Repeater (even for a simplex hotspot).

Name: anything you like. I used Pi-Star.

Sub Name: usually your country, but I used Hotspot.

Repeater callsign: enter your callsign and one or two spaces followed by a letter for the 'module.' A = 23cm band, B = 70cm band, C = 2m band. The module letter must be the 8th character.

Gateway callsign: enter your callsign and one or two spaces followed by a letter G for gateway. The module letter must be the 8th character.

Gateway IP address: is only used if you are setting up an Ethernet-connected D-Star repeater. Leave it blank.

Operating frequency: enter the transceiver receive frequency. i.e. the frequency that the repeater or hotspot is transmitting on.

Duplex: enter -DUP if you need the transceiver to transmit low. Enter +DUP if you need the transceiver to transmit high. Enter OFF if you are using a simplex hotspot. If there is no OFF option, enter -DUP or +DUP and set the offset to zero.

Offset Freq: enter the correct offset for your local band plan. Typically, ±600 kHz for a repeater or duplex hotspot on the 2m band, or ±5 MHz for a repeater or duplex hotspot on the 70cm band. Enter 0.000000 for a simplex hotspot.

Mode: The mode is DV for D-Star, FM for analog FM repeaters, or DD for the digital data mode on the 1.2 GHz (23cm) or higher band.

Tone and **Repeater Tone** are only used for FM channels. Enter the CTCSS tone and TONE or TSQL for FM repeaters.

Use (FROM): set to Yes.

Position: set to Exact or Approximate as appropriate. It does not matter.

Latitude and Longitude: in degrees, minutes, and seconds. Remember to add the N/S and E/W. If your radio has a GPS receiver you can get your position from that. Otherwise, Google Earth is a good resource. Set Tools > Options to Degrees, Minutes, Seconds. Zoom and right-click on your house. The latitude and longitude are indicated at the bottom of the window.

UTC Offset: set this to your current UTC offset.

Click on the Clone write icon (arrow pointing to handheld radio) to write the new radio configuration back to the radio.

Finally, click Save to save a backup on your PC.

Add your hotspot to the repeater list using the radio buttons

You cannot edit or add to the repeater list if the radio is in DR mode. Exit the DR mode back to the normal FM channel or VFO mode.

On the radio, click Menu > DV Memory > Repeater List. Select your country group. You can add a new group if you like but it is not necessary.

Touch and hold the top, or any line, and select Add.

Enter the required information (listed above). You will find that Type defaults to DV repeater and the Gateway callsign will auto-populate when you enter your callsign.

I do recommend filling in the latitude and longitude data because it will make your hotspot appear in the 'near repeater' list, saving you the bother of looking through the entire list for it.

Go down to the bottom of the list and select <<Add Write>>. You are done! It is not a bad idea to do a backup to your SD card at this stage. Menu > Set > SD card > Save Setting > <<New File>> > Ent > Yes.

Modifying the DV memory repeater list

Right-click the Repeater List folder and Export > All.

After I discovered that routing does not work on my hotspot or any repeaters in New Zealand, I modified the .csv data file to delete all of the other countries. I left the New Zealand repeaters and added my hotspot and the two Christchurch FM repeaters as discussed above.

When you have finished making changes, save the .csv file, and then import it back into the CS program. You must close the spreadsheet before importing it back into the CS program. Right-click the Repeater List folder and Import > All, then Clone Write the data back to the radio.

More CS program DV or DV/DD memory options

You can edit the repeater list group names in the Group Name list.

The DV or DV/DD memory sub-menu is also where you can add the reflectors you will use a lot into the Your Call Sign list. MENU > DV memory > Your Call Sign.

If you touch and hold a memory entry, you can add, edit, move, or delete DV or DV/DD memories. You can skip individual repeaters or the whole group from the DR scan. Confusingly, Skip all OFF includes the entire group in the scan, and Skip all ON skips the entire group.

Skipped repeaters cannot be selected by turning the Multi knob on a desktop radio, or the Dial knob on a handheld. They also fail to show up if you use the Near Repeater option. Your local D-Star repeaters, hotspot and any local FM repeaters should not be skipped.

If you have the full repeater list loaded, there is no point in scanning for repeaters in different countries or any repeaters that are out of the range of your radio. The only channels that you want to be included in the DR scan are your local repeaters. I found the easiest thing to do was to set Skip all ON to turn scanning off on all groups and then use the individual SKIP Soft Key to set my four local repeaters so they will not be skipped. Not that I actually use the DR scan anyway. To initiate a scan in the DR mode, press Menu > SCAN > Scan button to scan through your local channels. The scan can include FM repeaters if they are in the DV repeater memory bank.

"MY WAY"

This is now my default method for operating D-Star. The 'Your Call Sign' list is supposed to hold the callsigns of people you call often using the callsign routing method. The problem is that callsign routing does not work from a hotspot or on most repeaters. At least, I can't make it work and I have had advice from a local D-Star expert that it will only work between genuine Icom repeaters or if both stations are listening to the same repeater. But you can use the list to hold reflectors or even repeaters that you like to link to. Since the radio has no other memory for you to store your list of favourite reflectors, I think this is a good re-use of the 'Your Call Sign' option.

I used the CS software to load a list of reflectors and the unlink, information, and CQCQCQ commands. Now I can link to my favourite reflectors and change to CQCQCQ simply by turning the **Dial** or **Multi** knob.

Your Call Sign (Remain 189 memories)

No.	Name	Call Sign
1	Unlink	U
2	Echo	E
3	Info	I
4	Use Reflector	CQCQCQ
5	REF001 C	REF001CL
6	REF030 C	REF030CL
7	Kiwi chat	XLX299KL
8	WW 299	XLX299AL
9	BM530	XLX299BL
10	QuadNet Tech	XRF757CL
11	QuadNet Array	XRF757AL
New		

Figure 4: Re-purposing the Your Call Sign list

Create a list like the one above in the CS software. You can add as many reflectors and modules as you want. Save the file to your PC and upload it back to your radio as discussed above. Notice the L for linking at the end of the reflector callsigns.

To use the list, select the TO field using the up key or touching the TO icon on touchscreen radios. Press **Enter** or touch the text field and select **Your Call Sign**. Select a reflector and on a handheld, press **Enter** again. Transmit to establish the link. Then rotate **Dial** or **Multi** to the **Use Reflector** option before you make your call. You can use **Dial** or **Multi** to select a different reflector, unlink, echo, or information. "Neat!"

OPERATING MODES

You can operate D-Star in several ways.

1. Simply selecting DV (digital voice) means that the radio will transmit and receive using D-Star GMSK modulation. You can use this mode to talk to another radio directly in simplex mode or to talk to hams via a D-Star repeater or a hotspot. Channels can be stored on the standard memory banks alongside FM channels. The DV mode is not preferred for repeater and hotspot operation, but it will work.

2. Selecting the DR (digital repeater) mode is the preferred way to operate digital voice through a repeater or a hotspot. It is easier and provides more features.

3. D-Star radios that are equipped with the 23cm (1.2 GHz) and higher bands, support the DD (digital data) mode. This lets you send data such as computer files, websites, or photographs at a much higher speed. The IC-9700 includes the 23cm band and the new IC-905 extends D-Star operation to the 2.4GHz, 5.6 GHz, and optionally the 10 GHz band.

SIMPLEX (DV MODE)

You can call another radio directly. Simply select a digital simplex frequency (check your local band plan). Select the DV mode and call the other radio. It is not very common to use D-Star for simplex calls. Most people use FM for direct radio-to-radio conversations. You do not have to be registered on D-Star to make simplex calls.

LOCAL REPEATER (DV MODE)

I do not recommend using this method unless the repeater is <u>not</u> connected to the Internet. Use the DR 'Use Reflector CQCQCQ' mode instead.

You don't have to use the DR (digital repeater) mode to make a call over your local D-Star repeater to another local D-Star radio. You can save the repeater information in one of the standard memory channels in the same way that you would save an FM repeater channel. Select the channel. Select **DV mode** and make your call.

TIP: I think it is obvious, but for amateur radio newbies. NO! You cannot make a D-Star digital voice call over an FM repeater.

It is very important to note that you can only talk to **local** D-Star stations in the local repeater mode. You will not be able to reach any station that is arriving through a link to another repeater or reflector. This can be very confusing. You can hear them, but they can't hear you. Sometimes you can hear a call sign that you know is a local operator, but they may be accessing the repeater through an Internet connection.

They might be using a D-Star 'hotspot,' or they may be using a DMR or P25 radio into a repeater that is linked to your D-Star repeater.

TIP: You can make calls in the DV mode through your local repeater or hotspot to linked reflectors. You can use **Menu >CS** *and edit the R2 field to open a gateway and the UR field to establish a link to a reflector. Some people believe that this way is easier, but I believe that it is better to treat your hotspot as a repeater, add it to the repeater list, and use the DR mode.*

Generally, I think that it is best to treat all D-Star repeaters and hotspots as gateways and always use the DR (digital repeater) mode.

DIGITAL REPEATER (DR MODE)

Digital repeater is the normal operating mode for D-Star. It allows you to call anyone that you can hear on your local repeater, link the repeater or your hotspot to a reflector, or in some cases link the repeater or your hotspot to a remote repeater.

In other words, you can communicate with anyone on the worldwide D-Star network, plus dozens of reflectors that are linked to DMR talk groups and/or YSF rooms.

Use the 'Use Reflector' setting. The TO box will show an icon of a computer and 'Use Reflector CQCQCQ.' You do have to be registered to make DR (digital repeater) calls.

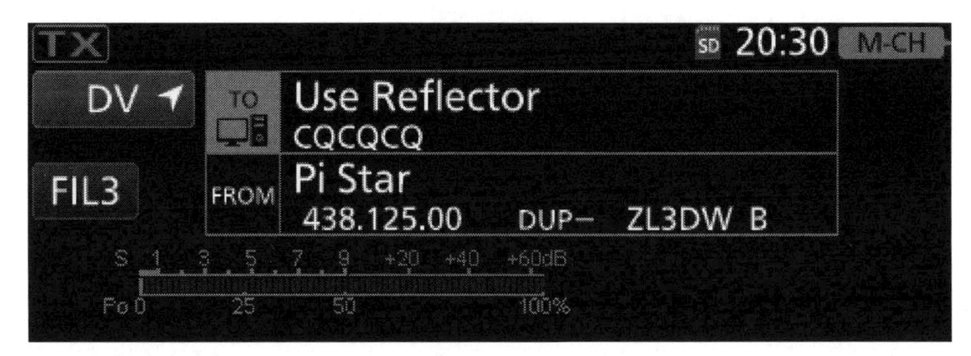

Figure 5: The DR mode

When you press and hold the DR key a new screen opens. On the IC-705 and IC-9700 and other touchscreen models, you can simply touch the text beside the TO and FROM boxes. On handheld radios without a touchscreen like the ID-51 and earlier models, use the **up** (RX->CS) or **down** (DR) pad keys to select TO or FROM and then click the centre 'Enter key.'

The TO box describes the call you are about to make when you press the PTT. If TO is highlighted by touching the TO icon or clicking the up button on a handheld, you can change the selection by turning the 'Dial' knob or 'Multi' on a desktop radio. If the radio is currently in the 'Reflector' mode, the choices will be, Use Reflector CQCQCQ, Link to Reflector, Unlink Reflector, Echo Test, and Repeater Information.

The FROM box carries information about the repeater or hotspot you are using, including the repeater name, its frequency and offset, and its callsign including a module letter. If FROM is highlighted by touching the FROM icon or clicking the **down** button on a handheld, you can change the selection by turning the 'Dial' knob or 'Multi' on a desktop radio. It will switch through your closest local repeaters and hotspots programmed into the DV memory.

Clicking the blue 'Enter' button when FROM is highlighted or touching the FROM icon and then the text beside the FROM box will pop up a 'From Select' menu. You can select a repeater from the Repeater list, a near repeater based on your GPS location, or one you have used before (FM or DV) from the TX History.

Clicking the blue 'Enter' button when TO is highlighted or touching the TO icon and then the text beside the TO box, will pop up a 'To Select' menu. This is where things get interesting! There are a lot of choices.

Local CQ: is used when you want to chat on your local repeater without the call being extended over a link to a reflector. This was discussed in 'Local Repeater (DV mode).' The local CQ mode is confusing and should be avoided.

The TO box will show CQCQCQ and the TO icon will show TO and three little people.

Selecting **Menu > CS** shows you the D-Star fields.

UR: CQCQCQ

R1: Your repeater or hotspot callsign e.g. ZL3DW B

R1: 'Not used' because there is no gateway link being established

MY: Your callsign/suffix

Gateway CQ: lets you select a repeater from the list in the DV memory. Once that has been selected, when you transmit, the radio will attempt to make a routed connection to the selected repeater. Unless you live in Japan, or both repeaters are genuine Icom repeaters, the call will fail to establish a link. Gateway routing does not work through most hotspots, or most (maybe all) non-Icom repeaters. Routing is very likely to fail, but some repeaters will accept a linked connection, see page 49.

Selecting **Menu > CS** shows you the D-Star fields.

UR: /5BACY C Troodos Mt. The repeater you are trying to route to

R1: Your repeater or hotspot e.g. ZL3DW B

R1: Your gateway link e.g. ZL3DW G

MY: Your callsign/suffix

Note that the / in the UR field indicates a routed connection.

Your Call Sign: is one of the odd things with D-Star. It does not hold your callsign, it is a favourites list. You can store reflectors that you call often in this list. It can be edited on the radio or in the Icom CS software. The list holds 200-300 entries depending on the radio.

Reflector: is where you want to be most of the time. The normal choice for talking on your hotspot or repeater is to select **Use Reflector**. This will set the TO icon to a computer and the text box will display 'Use Reflector CQCQCQ.' The Unlink Reflector, Repeater Information, and Echo Test are self-explanatory. Pages 41, 42, 44.

The 'Link to Reflector' option is important. Selecting it lets you select a previously used reflector from a list, or 'Direct Input.' Notice that the reflectors in the list all have an L at the end, indicating a linked connection.

The ID-51 and earlier radios only let you select from the REF reflectors. There is a way around that limitation. We will get to it later. The ID-52, IC-705, and IC-9700 let you change the reflector type to select REF, XRF, DCS, or XLX reflectors.

When you have chosen a reflector, you can transmit briefly to establish the link. You do not have to wait for the text to scroll across the screen. A couple of seconds is adequate. After the link is established, you will usually get a voice message and/or text scrolling to show the link. To avoid interrupting a conversation, this will not happen if someone is talking on the reflector. While linked, you will hear any conversations that occur on the reflector. Always change back to the Use Reflector CQCQCQ mode as soon as the link has been established. If you don't, the system will re-establish the link along with the voice message every time you transmit.

Selecting **Menu > CS** shows you the D-Star fields.

UR: REF001CL a link to module C on the REF001 reflector

R1: Your repeater or hotspot e.g. ZL3DW B

R1: Your gateway link e.g. ZL3DW G

MY: Your callsign/suffix

Note that the L at the end of the UR field indicates a linked connection.

RX History: is used to create a routed connection directly to another person. Icom says that it works, but because it uses routing, I have been unable to make this mode work from my hotspot or any of the three local D-Star and multi-mode repeaters. It will work if both callsigns are on the same repeater. When you transmit the receiving station gets a little bell alarm to say that you called. Plus, you get the usual callsign voice announcement so I'm not sure how much value there is in the alarm. Try it out. It might work on your repeater.

TX History holds a list of the last few connections you made, so it is very handy for reconnecting to a reflector you use quite frequently. I should probably use it more often, but I usually link using the Pi-Star dashboard.

Direct Input (UR) is a way of loading the UR field (TO box) directly. It is very handy if you want to link an ID-51 to a non REF reflector, and it is quicker than scrolling endlessly through the numbers on the 'Link to Reflector' page. Enter the reflector number, module, and L for link. For example, REF001CL or XLX299GL. Press the Enter key to load it into the TO box and PTT to establish the link. As usual, you should change back to the Use Reflector CQCQCQ mode as soon as the link has been established

Direct Input (RPT) establishes a routed call to a repeater. As explained earlier this does not usually work.

TIP for the IC-9700. There is a small "Gotcha" relating to the selection of the repeater that you are planning to use. The IC-9700 has two receivers but they can't be set to receive on the same band at the same time. For example, if you had the Main receiver set to the 2m band and the Sub receiver set to an FM repeater on the 70 cm band. When you select the DR mode and touch the 'FROM' icon on the Main receiver, you will not be able to set it to any 70 cm repeaters because the Sub receiver is already using the 70 cm receiver.

You can put both receivers into DR mode and listen to DV repeaters on two different bands at the same time. Or you can have one band on DV and the other on FM. But you cannot listen to two repeaters on the same band at the same time. You get a better display if you use the main channel for the DR mode.

TIP: If you are using a hotspot there are two additional ways to link to a reflector. You can select a reflector on the Admin tab of the Pi-Star dashboard, or you can use the Apple iOS or Android version of 'ircdbremote.'

TIP: Bluetooth-capable models like the IC-705 and ID-52A can use the RS-MS1A App for Android or iOS for a wide variety of functions including linking to reflectors and gateway repeaters.

The DR mode

The previous chapter is enough to get you to the point where you can access a repeater and make a call. This chapter covers the DR mode in more detail. The following chapters fill in more information about D-Star and how to build and configure a Pi-Star hotspot.

CONNECTING TO A REFLECTOR

IC-705, IC-9700 and similar radios

The link is achieved using the 'Link to Reflector' option. Double tap the text to the right of the **TO** icon, to show the 'TO SELECT' sub-menu and select **Reflector > Link to Reflector**.

At this stage, the menu will show **Direct Input** and it may list some previously selected reflectors. You can select the one you want if it is listed. Assuming that you are starting from scratch, select **Direct Input** to get to the Link to Reflector sub-menu.

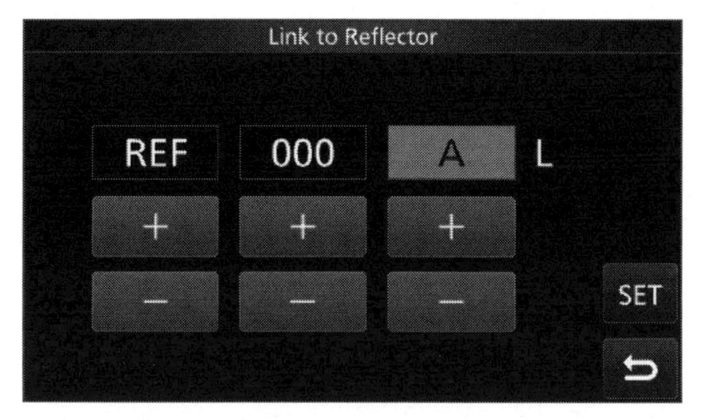

Link to Reflector sub-menu

Use the plus and minus keys to select REF, XRF, DCS, or XLS. Then the reflector number, and the module (band) letter. Then touch SET. The sub-menu will close, and you should see the correct 'Link to Reflector' text next to the 'To' icon. In the example above, it would be REF000AL.

Before connecting to a reflector, make sure that the repeater is not already linked to another reflector. Before you unlink the repeater, put a call out and let the other repeater users know that you are planning to unlink the repeater. Someone may be waiting for a call, or for a Net to start. It is good etiquette to restore the linking after you have finished using the repeater.

Check the repeater dashboard on your PC to see where the repeater is linked or send an Information request on the radio and take note of the linking information.

To connect to the reflector, press the Microphone PTT for a second or two. There is no need to wait for the characters to scroll through. You may get a scrolling text message and often a voice message stating that the repeater is now connected to the reflector.

Once the connection has been made you should immediately use the **Multi** knob to return to **Use Reflector CQCQCQ**. You can also do this by touching the **TO** icon if it is not already highlighted and selecting **Reflector > Link to Reflector > Use Reflector CQCQCQ**.

TIP: On radios, with a VFO knob you can select **TO** *and rotate the VFO until the display shows* **Use Reflector CQCQCQ**. *But I think the Multi knob is the most intuitive method.*

Listen for traffic on the reflector before calling CQ or answering another station.

ID-52, ID-51, and similar radios

Press the **up** (RX-CS) button and **Enter** (centre dot button) to highlight the **TO** icon.

You can also select a reflector from the list of recently used reflectors. Select **Reflector > Link to Reflector** and choose one of the reflectors from the list.

Or use **Reflector > Link to Reflector > Direct Input** and use the **Dial** and **right** (CS) button to select a REF reflector then press the **Enter** button.

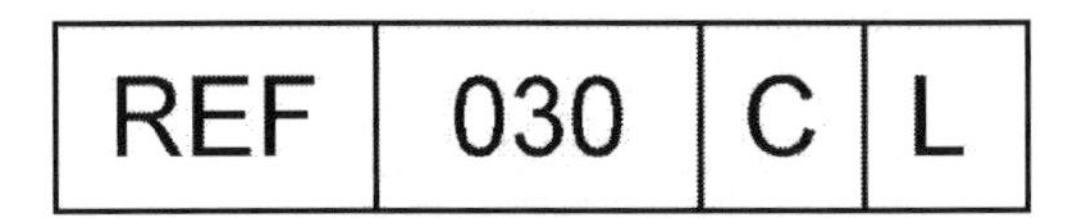

REF	030	C	L

- Select reflector type (ID-52 only), then press the **right** (CS) button.
- Select the reflector number, then press the **right** (CS) button.
- Select the reflector module, then press the **Enter** button.

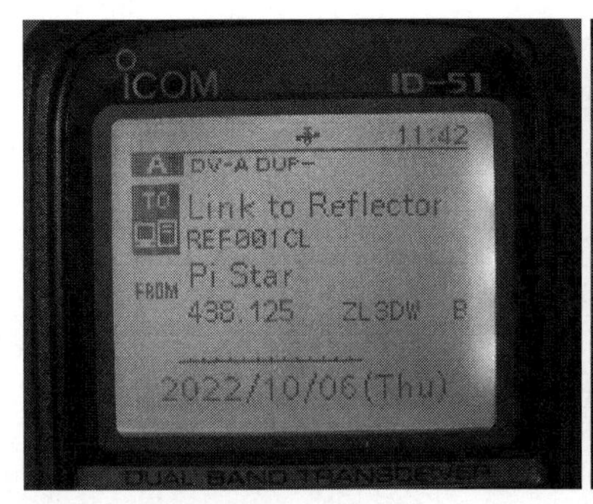

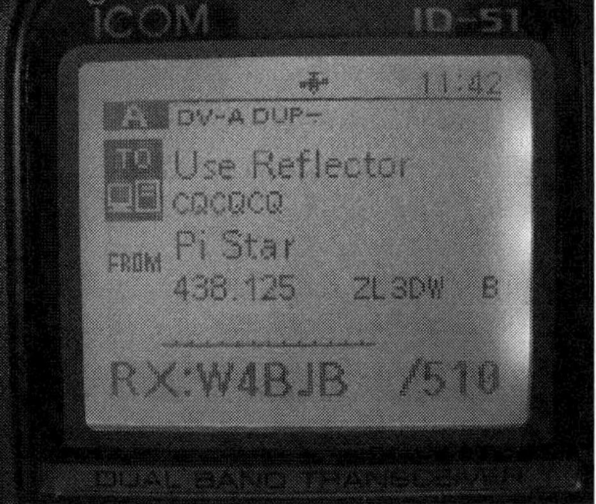

Figure 6: Linking to a reflector on the ID-51

TIP: Using this method, the ID-51 can only select REF reflectors. You can select XLX, XRF, and DCS reflectors by entering them into the **Menu > Direct Input (UR)** *submenu. Enter the reflector number, module, and L for link. Then press* **Enter**. *For example, XLX299KL. See Advanced linking and routing for the details.*

Once the reflector has been selected, briefly press the PTT, to send the request through the network. Then use the **Dial** knob to change to 'Use Reflector CQCQCQ' mode.

Pi-Star linking to a reflector or a repeater or hotspot node

If you are using a hotspot, you can create a link to a reflector using the 'D-Star link Manager' on the Pi-Star dashboard 'Admin' tab. Just choose the reflector and module with the drop-down lists and click **Request Change**. Your radio should respond with a text and voice announcement and the Pi-Star dashboard should indicate that it is linked. If it does not link, the channel is probably busy, wait 30 seconds then try again.

D-Star Link Information

Radio	Default	Auto	Timer	Link	Linked to	Mode	Direction	Last Change (NZST)
ZL3DW B	REF001 C	Auto	Never	Up	REF001 C	DPlus	Outgoing	17:23:54 Sep 17th

D-Star Link Manager

Radio Module	Reflector	Link / Un-Link	Action
ZL3DW B	REF001 C	Link UnLink	Request Change

Alternatively, you can select **Text Entry** in the reflector drop-down list and type in the node callsign. Select the correct module for the node and click **Request Change**.

To unlink a reflector or node, select **Unlink** and click **Request Change**.

Phone app linking to a reflector

The 'ircDDB Remote' Android or Apple iOS phone apps written and developed by David Grootendorst PA7LIM control Pi-Star hotspots over WiFi. Make a selection or save the reflectors you use often.

Icom's RS-MS1A software is an Android phone app that lets you control DR mode linking and unlinking of reflectors from your phone, and a lot more besides that. The process is the same as linking to a reflector on the radio, but of course, the phone has a touchscreen. Most radio models require a data cable, but radios equipped with Bluetooth like the ID-52 and IC705 can use that. The RS-MS1I app for Apple iOS phones and iPads only works with Bluetooth-capable D-Star radios.

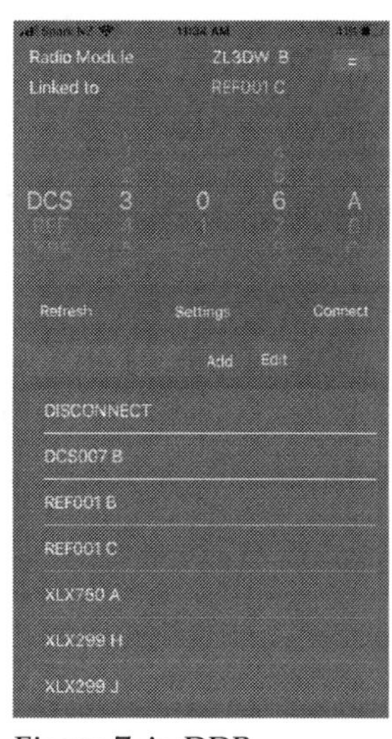

Figure 7: ircDDB

D-STAR FIELD STRUCTURE

D-Star uses four special data fields which are embedded into all transmissions. They are UR, R1, R2, and MY. You can see the contents of these four data fields if you select MENU > CS on a handheld radio, or Menu > 2 > CS on the IC705 and IC-9700.

UR, 'Your Call' holds the information being sent to the repeater and the Internet. It is the information displayed in the TO box. For normal repeater or hotspot operation, it will be set to Use Reflector CQCQCQ.

This could have been labelled 'TALK,' but I guess CQ is more universally understood. The data in the UR field will change if you are sending a signal to obtain information or link to a repeater or reflector.

- In the table below '_' indicates a space.
- CS indication is the information that you will see on the CS menu screen MENU > CS
- TO text box is the information that you will see in the TO box in DR mode.

CS indication	TO text box	Function
CQCQCQ	CQCQCQ	Talk on a local repeater
CQCQCQ	Use Reflector CQCQCQ	Talk on a local repeater or through a gateway to a reflector or a linked repeater
_ _ _ _ _ _ _ I	Repeater Information I	Ask the repeater for information and check if the repeater is linked.
_ _ _ _ _ _ _ U	U	Unlink the repeater so that you can link to a different repeater or reflector or make a local repeater call.
_ _ _ _ _ _ _ E	Echo Test E	Echo can be used on some repeaters. It will send back a few seconds of your own transmitted audio, as a test signal.
ZL2RO	ZL2RO	Call Sign routing to a particular station, for example, ZL2RO. Unlikely to work.
REF001CL	Link to Reflector REF001CL	Connect to the REF001C reflector. The L indicates a link.
/VK8RADC	Darwin VK8RAD C	Gateway (routed) connection to module C on the VK8RAD 2m repeater in Darwin Australia. Unlikely to work.
VK8RADCL	Darwin VK8RADCL	Link to module C on the VK8RAD 2m repeater in Darwin.

R1, 'Repeater 1' is set to the call sign of the repeater or hotspot that you are calling through. It is the information displayed in the **FROM** box. The call sign is followed by a module letter which must be the eighth character. For example, 'ZL3DV_ _ B' would be a 70 cm repeater in New Zealand.

If R1 is blank or (NOT USED*) it means that you are making a Simplex call. No local repeater is selected.

Module	Function
A	23 cm repeater
B	70 cm repeater
C	2 m repeater
D	Digital Data

TIP: Japan uses A for 70 cm repeaters and B for 23 cm repeaters. Some systems have multiple modules on the same band, so B, C, and D might all be 70 cm repeaters. It does not really matter so long as you have the correct module code for the repeater you will be accessing with your radio. If the repeater call sign is correct but the letter is incorrect you will not get a message back when you transmit. In that case, try a different letter.

R2, 'Repeater 2' is set to the call sign of the gateway. The call sign is always followed by the letter G, for gateway. The G must be the eighth character. For example, 'ZL3DV_ _ G' would be the gateway for the ZL3DV repeater.

Module	Function
G	Gateway

If R2 is blank or (NOT USED*) it means that you are making a Local Repeater Call, otherwise known as a Local CQ call. No gateway repeater is selected.

MY, 'My Call' holds your call sign including the four-character suffix that you set in the First Steps above. The field holds eight characters before the / symbol and four characters after the / symbol.

You can change from one version of your callsign to another, or edit the data that will be sent, in the **CS sub-menu**.

Or the **Menu** > Set > **My Station** > **My Call Sign (DV)** setup.

Callsign (8) /Suffix (4)
ZL3DW /9700

MONITORING A DV REPEATER OR HOTSPOT (DR MODE)

Press and hold the DR button to put the radio into digital repeater mode. The TO box will probably show 'CQCQCQ' or 'Use Reflector CQCQCQ.'

We might as well set the radio so that it is ready for transmitting.

IC-705, IC-9700 and similar radios

Touch the text to the right of the TO icon (possibly twice) to open the TO SELECT sub-menu. Select Reflector > Use Reflector. The text next to the TO box will show Use Reflector CQCQCQ, with an icon showing a computer and screen.

ID-52, ID-51, and similar radios

Press the up (RX-CS) button and Enter (centre dot button) to highlight the TO icon. Select Reflector > Use Reflector the TO box will show Use Reflector CQCQCQ.

Next, you have to set the radio to receive signals from a repeater near you. This could be a D-Star or multi-mode digital repeater at a high site, a 'simplex repeater' or high-power hotspot hosted at someone's house, or a hotspot connected to your PC.

IC-705, IC-9700 and similar radios

Touch the text beside the FROM icon. You may have to touch it twice, to bring up the FROM SELECT sub-menu. You will see three menu options.

- Repeater List – lets you select a repeater from your DV/DD repeater list. Note that it must be a repeater that is near you. If you are not within the repeater's RF coverage area it won't work.
- Near Repeater - lets you select from a list of repeaters that are located near you. This will only work for your hotspot and local repeaters if you have their GPS location included on the DV memory repeater list entry.
- TX History – lets you select from a list of recently accessed repeaters. Note that recently accessed FM repeaters will not be included unless you were using the DR mode at the time. If you are regularly accessing two or three local repeaters, the TX History (DV) option is probably the easiest way to switch between them.

ID-52, ID-51, and similar radios

Press the down (DR) button and Enter (centre dot button) to highlight the FROM icon. Select Repeater List > Choose a Repeater Group > Choose a Repeater. The FROM box will show the repeater name, frequency and callsign.

You can also change the FROM data using the Dial, searching Near Repeater, or choosing from TX History.

Listen to the repeater or hotspot

After you have selected your local repeater or hotspot, you should have some information in the **FROM** text box and you will hear anyone using the repeater… probably.

There is a potential "Gotcha" here! You will only hear D-Star traffic on the repeater. Some repeaters are capable of multi-mode operation. If you see an indication on the S meter but don't hear any audio, the chances are that a user is using the repeater for the DMR, P25, or digital transmission mode. There are also private squelch options such as D.SQL and CSQL, but you should not have those active at this stage.

The **FROM** box will display the name of the repeater, its frequency, DUP+ or DUP- duplex setting, and the callsign of the repeater with an A, B, or C letter indicating the band. The frequency will change to indicate your transmit frequency while you are transmitting.

If you select an FM repeater stored in the DV/DD memory list, the radio will switch to FM and the text will display similar information without the repeater callsign.

TIP: A simplex repeater or hotspot will also show DUP+ or DUP- because the radio does not understand the concept of simplex repeaters. Don't be concerned, the duplex offset will be set to 0.000 so the transceiver will transmit and receive on the same frequency.

CHECK THE REPEATER LINK STATUS

Before you link or unlink a repeater, it is a good idea to find out if, and where, it is currently linked. There are three ways to find this out.

The repeater will occasionally announce the link connection, either by a message scrolling across the screen or by a voice announcement. But you probably don't want to wait for that to happen.

You can go onto the repeater's dashboard page on the Internet or your Pi-Star dashboard if you are using a hotspot to see what linking is in place.

Or you can transmit the 'I' code to the repeater which will respond with the information.

Return to ***Use Reflector CQCQCQ*** *mode after you have received the information, or every call you make will create an Information report from the repeater.*

IC-705, IC-9700 and similar radios - using the I command

Touch the TO text box. You may have to touch it twice.

Select Reflector > Repeater Information. The TO box will change to Repeater Information I. Key the Microphone PTT for a second or two. You do not have to wait for the text to scroll.

You will see a message scroll across the screen and in most cases a voice message saying that the repeater is linked to a repeater or reflector, or that it is "not linked."

IC-705, IC-9700 and similar radios - using the DTMF 0 command

The DTMF codes are primarily for people using handheld radios where a DTMF code can be sent with a single button press. Unless you are using a microphone with a DTMF keypad, it is a little more difficult on the IC-9700.

Press MENU > 2 > DTMF > SEND > Direct Input > 0 > TX.

To make this easier, you can store commonly used DTMF codes using the EDIT key. That will shorten the sequence slightly to MENU > 2 > DTMF > SEND > d0:0. See DTMF codes on page 46.

ID-52, ID-51, and similar radios

Press the up (RX-CS) button and Enter (centre dot button) to highlight the TO icon. Select Reflector > Repeater Information, the TO box will show Repeater Information I. Key the transmitter for a second or two then release the PTT and listen to the repeater ID message.

UNLINK A REPEATER

It is good manners to put a call out to let the other repeater users know that you are planning to unlink the repeater. Someone may be waiting for a call from a linked reflector or for a Net to begin. It is also good etiquette to restore the linking after you have finished using the repeater. Check the repeater dashboard on your PC to see where the repeater is linked. Or send an Information request on the radio and take note of the linking information.

Of course, this is not necessary if you are unlinking your own hotspot.

TIP: Most references including the Icom manuals suggest that you should unlink a repeater before linking to another reflector. In practice, this seems to be unnecessary.

To unlink a repeater, you can use the menu structure in the radio to send the U command, or on many repeaters, you can send a DTMF command.

*Return to the **CQCQCQ** mode after you have unlinked the repeater, or every call you make will create an Unlink report from the repeater.*

IC-705, IC-9700 and similar radios - using the U command

Touch the TO text box. You may have to touch it twice.

Select Reflector > Unlink Reflector. The TO box will change to Unlink Reflector U.

You can use this command to unlink from a gateway repeater as well.

Key the Microphone PTT for about a second. You do not have to wait for the text to scroll. You will see a message scroll across the screen and in most cases a voice message saying that the repeater is now "not linked."

Now that you have unlinked the repeater, you can link to another repeater or reflector, or you can make a call out to the local area.

Touch the TO box again and select Local CQ if you want to make a local call, Gateway CQ if you want to select a link to a repeater, or Reflector if you want to connect to a reflector.

IC-705, IC-9700 and similar radios - using the # DTMF command.

The DTMF codes are primarily for people using handheld radios where a DTMF code can be sent with a single button press. Unless you are using a microphone with a DTMF keypad, it is a little more difficult on the IC-9700.

Press MENU > 2 > DTMF > SEND > Direct Input > # > TX.

To make this slightly easier, you can store commonly used DTMF codes using the EDIT key. That will shorten the sequence to MENU > 2 > DTMF > SEND > d1:#. See DTMF codes on page 46.

*TIP: Sending a ** DTMF code instead of a # code will reset most repeaters to their default link. This is a handy way of setting the repeater back to 'normal.' MENU > 2 > DTMF > SEND > d2:**.*

ID-52, ID-51, and similar radios

Press the up (RX-CS) button and Enter (centre dot button) to highlight the TO icon. Select Reflector > Unlink Reflector. The TO box will show Unlink Reflector U. Key the transmitter for a second or two then release the PTT and listen to the repeater "not linked" audio message.

Pi-Star hotspot

If you are using a hotspot, you can unlink a reflector or a linked repeater using the D-Star Link Manager on the Admin page of the Pi-Star dashboard. Just select the Unlink bullet point and click Request Change.

Tip: You do not have to unlink a reflector before selecting another one.

THE ECHO FUNCTION

The Echo function works on most repeaters. The function records and repeats back a few seconds of your transmission so that you can confirm that you are connected and hear what your signal sounds like.

After you have carried out an Echo test, return to CQCQCQ mode. Don't leave the radio in Echo mode or every call you make, will be repeated back to you.

IC-705, IC-9700 and similar radios

Touch the TO text box. You may have to touch it twice.

Select Reflector > Echo test. The TO box will change to Echo Test E.

Key the Microphone PTT for about a second and make a short 5-second announcement such as "*K3BLAH testing on Repeater K3AA*." You do not have to wait for the text to scroll.

If the function is enabled, you will hear your transmission repeated back to you.

After you have made the Echo call you must always touch the TO box again and select Local CQ if you want to make a local call, Gateway CQ if you want to select a link to a repeater, or Reflector if you want to connect to a reflector.

ID-52, ID-51, and similar radios

Press the up (RX-CS) button and Enter (centre dot button) to highlight the TO icon. Select Reflector > Echo Test, the TO box will show Echo Test E. Key the Microphone PTT for about a second and make a short 5-second announcement such as "*K3BLAH testing on Repeater K3AA*" then listen for the message to be repeated back.

MAKING A LOCAL CALL IN DR MODE

If you want to make a call to your local area and you don't want the world listening, you can use the Local CQ mode. Before you unlink the repeater, put a call out and let the other repeater users know that you are planning to unlink the repeater. Someone may be waiting for a call from a linked repeater or for a Net to start on a reflector. It is good etiquette to restore the linking after you have finished using the repeater.

So, check the repeater dashboard on your PC to see where the repeater is linked, or send an Information request on the radio and take note of the linking information.

IC-705, IC-9700 and similar radios

Touch the text beside the TO icon (possibly twice) to open the TO SELECT sub-menu. Select Local CQ the text next to the TO box will show CQCQCQ and the TO icon will include three little people. Then make a standard CQ call or answer another station the way you would on an FM repeater.

ID-52, ID-51, and similar radios

Press the up (RX-CS) button and Enter (centre dot button) to highlight the TO icon. Select Local CQ. The 'TO' box will show 'CQCQCQ.' Then make a standard CQ call or answer another station the way you would on an FM repeater.

Note

In this mode, you can only transmit through your local repeater to D-Star stations within the repeater's coverage area. You cannot talk to anyone accessing your repeater through a gateway or reflector. You can hear them, but they can't hear you. Generally, unless the repeater is not capable of being linked, it is best to use the 'Use Reflector CQCQCQ' mode.

MAKING A CQ CALL ON A LINKED REPEATER OR HOTSPOT

It is always best to use this method. You will transmit through your local repeater to D-Star stations within the repeater's coverage area and you can also talk to anyone accessing your repeater through a gateway or reflector.

IC-705, IC-9700 and similar radios

Touch the text beside the TO icon (possibly twice) to open the TO SELECT sub-menu. Select Reflector > Use Reflector the text next to the TO box will show Use Reflector CQCQCQ and the TO icon will include three little people. Then make a standard CQ call or answer another station the way you would on an FM repeater.

ID-52, ID-51, and similar radios

Press the up (RX-CS) button and Enter (centre dot button) to highlight the TO icon. Select Reflector > Use Reflector. The TO box will show Use Reflector CQCQCQ. Then make a standard CQ call or answer another station the way you would on an FM repeater.

Note that if the repeater is a hotspot or simplex receiver, you always have to use this 'Use Reflector CQCQCQ' mode. The Local CQ mode will not work at all on a simplex repeater.

MULTI-MODE DIGITAL REPEATERS

Some repeaters are capable of transponding multiple digital modes, often D-Star, P25, DMR, and Fusion. Sometimes FM as well. Special rules apply to the use of these repeaters because users of other equipment will not be able to hear your transmissions and you will not be able to hear theirs. You need to be very aware of the status of the repeater. If you are at home or have an internet connection this is usually done by observing the repeaters 'dashboard' website. The dashboard will show who is using the repeater and the mode that is currently in use.

If the repeater is being used for a different digital mode, you will be able to see a signal on the radio S meter, but you won't be able to hear it. Usually, there is a timer to stop other signals from interfering with an ongoing QSO. You can't use the repeater for a different digital mode until the repeater has been clear for several seconds.

USING A SIMPLEX REPEATER

If the repeater you are using is a hotspot or simplex repeater, you must use the 'Use Reflector CQCQCQ' mode. The 'Local CQ' mode will not work on a simplex repeater and your transmission will not be broadcast locally. A Simplex repeater must always be linked somewhere. In this mode, you will transmit through the repeater or hotspot to anyone accessing your repeater through a gateway or reflector. But you cannot talk to local stations unless they are using a different repeater or a hotspot that is linked in via the gateway.

DTMF CODES

Many repeaters can use short DTMF codes to control repeater functions including linking to reflectors. I think that it is more trouble than it's worth.

IC-705, IC-9700 and similar radios

On the IC-705, IC-9700 and similar radios, you can send a DTMF code by using the DTMF screen. MENU > 2 > DTMF.

Touch SEND to immediately send a code using Direct Input, or you can send one of the last codes sent. Touch EDIT to pre-load up to 16 saved codes. SET changes the speed that the codes are sent from the transmitter. I have changed to 200ms because my repeater seemed to miss the codes when the speed was set to 100ms. But you can try 100ms first. Higher numbers send the codes more slowly.

ID-52, ID-51, and similar radios

You can send DTMF codes from the DTMF memory or directly enter them. The ID-52 can store 16 DTMF code strings of up to 24 characters.

Use QUICK > DTMF TX > and select Direct Entry or d0 .. d16. The code that will be sent is beside the memory number.

You can also send a saved DTMF code using MENU > SET > DTMF/T-CALL > DTMF Memory > d0 .. d16. Press MENU to exit. Press the PTT to transmit the code.

To store a DTMF code to the radio, use MENU > SET > DTMF/T-CALL > DTMF Memory > select a memory slot to use but do NOT press Enter. Press QUICK > Edit > enter the DTMF code then push Enter.

Common DTMF repeater codes.

Unlink repeater	#	(same as UR = _ _ _ _ _ _ _ U)
Linking information	0	(same as UR = _ _ _ _ _ _ _ I)
Reset to default link	**	(resets the gateway to its default linking)

Some repeaters support DTMF gateway linking to reflectors:

(example) Link to a DCS reflector DCS001A	D1A or D101
(example) Link to a DCS reflector DCS005B	D5B or D502
(example) Link to a REF reflector REF006C	*6C
(example) Link to a REF reflector REF005A	*5A

It seems that you can only access reflectors from 1 to 9 using DTMF since the last two numerals are used to select the module letter. The letters A to Z are represented by numbers 01 to 26. For example, to access DCS006Q you would send a DTMF code of D617.

D-STAR (DR) SCAN MODE

The Icom radios can scan the VFO, programmed scans between two frequencies, the FM or DV channels stored in the main memory banks, or if you are in the DR mode, the repeaters in the DV repeater list. The Scan button enables the scan mode appropriate to the mode that the radio is in. In the VFO mode, you get a choice of scanning all frequencies, the current band, the linked program banks, or the programmed scan ranges. You can also scan on TONE. In this mode, the radio stays on the frequency and rolls through all the CTCSS tones. It is used to identify the tone being transmitted by a repeater, or possibly on the repeater input frequency. The repeater usually transmits the same tone as it requires you to send to open its squelch.

IC-705, IC-9700, and similar radios

Touch the FROM icon and you can manually step through the not skipped DV repeater channels using the MULTI knob.

To start a scan, press and hold the Scan button. You can select from Normal, Near Repeater All, Near Repeater DV, or Near Repeater FM. 'Normal' scans all repeaters that are not set to skip. The three 'near repeater' options scan the 20 closest DV repeaters and/or the 20 closest FM repeaters to your current location. Repeaters and banks marked 'skip' will not be scanned.

Press the Scan button, or press the PTT to stop the scan.

You only want to scan the local channels and hotspots within the range of your radio. The ones you can trigger. All other channels and overseas groups should be excluded from the scan using the 'skip' setting. There is no point in listening for channels that are too far away to be received.

I found the easiest thing to do was to set 'Skip all ON' on all of the groups and then use the individual SKIP Soft Key to set my four local repeaters so they will not be skipped. To change the Skip settings, open the DV or DV/DD memory bank using **MENU > 2 > DV MEMORY > Repeater List**.

Touch and hold a memory group title and select **SKIP ALL ON** to skip the entire group. When you have blocked all of the groups, touch your local group and touch and hold the entry for your local D-Star repeater. On the quick menu, touch **SKIP** to 'un-skip' that channel. It should now be the only channel that does not have 'SKIP' on the right side of the listing. Repeat this for any other local repeaters.

ID-52, ID-51, and similar radios

Click the **down** (DR) pad and you can manually step through the not skipped DV repeater channels using the **DIAL** knob. Or press and hold the **Mode/Scan** button to start the DR scan. You can select from **Normal, Near Repeater All, Near Repeater DV**, or **Near Repeater FM**. 'Normal' scans all repeaters that are not set to skip. The 'near repeater' options scan the 20 closest DV repeaters and/or the 20 closest FM repeaters to your current location. Repeaters and banks marked 'skip' will not be scanned.

Press the **Mode/Scan** button, or press the PTT to stop the scan.

The scan is very fast. It will pause on active channels. If you have the DV memory full of repeaters and you have not set the banks to 'skip,' the radio will scan repeaters in different countries. This is a waste of time since the radio can only reach channels that are close to your location.

CROSS MODE OPERATION

Often a reflector that you link to will have connections into other D-Star repeaters or reflectors and other digital voice mode rooms or talk groups. You can make calls to stations that are listening or calling via the distant reflector irrespective of the digital mode that they are using. For example, if you transmit a D-Star signal into your local D-Star repeater and it is linked either directly or through a reflector to a DMR repeater, you can talk to somebody who is using a DMR radio.

The reflector you are using might be connected to a Brandmeister, DMR+, or TGIF talk group, a talk group on a regional DMR network, or a combination. Some reflectors have modules connected to P25, NXDN, or YSF rooms. And some connections are to multi-mode groups such as the QuadNet hubs.

Advanced linking and routing

Routing should be possible, but it has fallen out of popularity, and I had trouble making sense of it. Almost everybody uses linking connections. In fact, most people only link to reflectors.

REFLECTOR LINKING

The normal method of linking to a reflector is covered in the 'Use Reflector' and 'Link to Reflector' sections in the last chapter. But you can also perform linking without going through the menu structure. I find using the menu more intuitive, but this way can create a shortcut and save it in the 'Your Call Sign' list. Touch the text next to the TO box > **Your Call Sign** > then touch and hold the top, or any, used line and select **Add or Edit**. Note you cannot edit an entry that is currently selected in the TO box. It gets me every time!

To enter a direct link. Touch on the text next to the TO box. **Select Direct Input (UR)**. Enter the direct link. An **L** in the 8th position indicates that you are requesting a link.

For example, entering **REF001CL** into the UR field (TO box) and briefly transmitting will link your transceiver to reflector REF001 C. Remember to change to **Use Reflector CQCQCQ** to talk after you have established the link. Use the **U** command in the 8th position, to unlink the reflector connection.

TIP: Some repeaters do not allow user linking to reflectors. They are permanently associated with a specific reflector or gateway node.

This method has the same result as selecting the reflector through the radio menu structure or a phone application, or if you are using a hotspot, linking to a reflector via the Pi-Star Admin page.

GATEWAY LINKING

Gateway linking links your repeater or hotspot to another repeater or hotspot. This mode is rarely used, and it only works on some repeaters. In New Zealand, only ZL1ZLD will accept a gateway link. It will not accept a routed gateway connection.

When you have made a gateway link you will hear any traffic on the linked repeater node and any reflector that the node is currently linked to. This could be handy if you wanted to monitor the reflector while also listening for a friend who uses that repeater. Otherwise, you might just as well link directly to the reflector.

The downside when you are linked directly to a repeater is that you cannot change the reflector that the repeater is linked to. If you attempt to do that you will connect directly to that reflector and lose the gateway link.

Direct links using the UR field (TO box).

To enter a direct link. Touch on the text next to the TO box. **Select Direct Input (UR)** and enter the direct link with an **L** in the 8th position. For example, entering **ZL1ZLDBL** into the UR field links your transceiver to module B on the ZL1ZLD repeater. Remember to change to **Use Reflector CQCQCQ** to talk after you have established the gateway link. Otherwise, the network will try to re-establish the link every time you transmit. Use the **U** command in the 8th position, to unlink the gateway.

Linking is different to routing. After you establish the link, you can talk via the **Use Reflector CQCQCQ mode**. Any station on the other repeater can talk back to you. They do not have to establish a route back to your repeater the way they would have to with a routed connection.

Pi-Star linking to a gateway repeater

To link to a gateway repeater, select **Text Entry** in the reflector drop-down list and type in the node callsign. Select the correct module for the node and click **Request Change**. For example, enter **VK2RAG**, select **C**, and click the button. You will hear any traffic on the VK2RAG C repeater node and any reflector that the node is currently linked to. To unlink a reflector or node, select **Unlink** and click **Request Change**.

Hotspot-to-hotspot linking

You can use gateway linking to link your hotspot to a friend's hotspot. Unless someone is close enough to receive the hotspot transmitter or your radio, this creates a private connection. I have not tried this but there is a video by David Capello KG5EIU, at https://www.youtube.com/watch?v=g28xAKt3Vkg&t=52s, and one by and one by K9WLW at https://www.youtube.com/watch?v=CkY9PrroE58.

Success when attempting to make a direct link to a hotspot, depends on the hotspot. Some will accept an incoming link and others will not. Pi-Star hotspots will accept an incoming connection, although this may depend on your firewall settings. According to at least one online source, Open Spot hotspots will not accept an incoming call, but they can link to a Pi-Star hotspot. The process is the same as linking to a gateway. Any connected hotspots will show on the 'D-Star Link Information' section on the 'Admin' page of the Pi-Star dashboard. They should be able to talk to you, and each other.

LINKING NOTES

If you or someone else has already established a link, you can use it by setting your radio to the **Use Reflector CQCQCQ mode**.

If you did not establish the link, it is good manners to put out a call to ask if it is OK to unlink the repeater before linking it to another reflector (or gateway node). Someone could be waiting for a friend or a net to come up on the channel.

A stacked repeater cannot have two modules linked to the same destination. For example, ZL3DVR A and ZL3DVR B cannot both be linked to the same reflector. This is one reason that user linking may be unavailable.

CALLSIGN ROUTING

Unfortunately, callsign routing will not work through a hotspot or a non-Icom repeater that is using a third-party repeater board. I have never been able to get callsign routing to work over the D-Star network between two local repeaters or between a repeater and my hotspot. It works fine if you are both using the same repeater. You get a little bell noise on the radio if the other radio calls you. You can route the call back by putting the other person's callsign into the UR box on your radio. If you are both on the same repeater you can also use the CQCQCQ mode.

Callsign routing is intended to be a way you can call a specific callsign, even if you do not know what repeater they are listening to. For example, entering **VK1ABC** into the UR field (TO box) is supposed to route the call to that specific callsign. The D-Star system routes the call to the last heard repeater. If it has no 'last heard' record, it routes the call to the station's home repeater. The one they registered for D-Star on. Note that this is not a private call. Both sides of the conversation will be heard on both connected repeaters.

Callsign routing seems like a great idea. But it fell into disuse because if you establish a link to another callsign you still take up the resource of both repeaters. The routed radios can only hear the other routed radio. Everyone else on the repeater can hear them but cannot talk back to them. This is not a great situation. It causes frustration and is not a popular option. Theoretically, if you enable the **Callsign Routing** option in the D-Star configuration, you can use callsign routing hotspot to hotspot. In that situation, you will not bother other users, so it is OK to use it. I have not been able to test this. I cannot get callsign routing to work except when both radios are listening to the same repeater.

GATEWAY CQ - NODE ROUTING

Selecting a repeater using the Gateway CQ method in the radio attempts to establish a route to the gateway node, not a link. This is very unlikely to work. If you look at the D-Star fields, you can see that it is a routed connection. See **Menu > CS**, or **Menu > Call Sign** on the ID-51 and ID-52. The UR field starts with a / indicating a routing connection. The mode might work between genuine Icom repeaters. But it has never worked for me.

The DV mode

I stated earlier that I believe that it is best to operate D-Star using the DR mode where the D-Star fields are automatically filled in for you. However, some people prefer to use the DV mode. So, I will cover that method and you can choose the way that suits you best. It's all about the way that the linking information gets into the four D-Star fields.

PRELIMINARIES

If you want to use D-Star reflectors or link to gateway repeaters, you still have to register your callsign. That was covered back on page 22.

SAVING THE CHANNELS

You can set the channels manually, but it is preferable to load channels into the standard memory bank using the CS software. You can create channels for your repeater or hotspot in much the same way as you would for a DMR radio. This is essentially the same as "my method," using the 'Your Call Sign' list and selecting from that in DR mode using the Dial or Multi knob. Except, in this case, you lose the functionality of selecting the repeater and hotspot using the FROM control. Annoyingly, if you use this DV method, you have to repeat all of the channels for all of the repeaters that you want to use. Just like you do on a DMR radio. It is clumsy but it works. Here you can see channels for my hotspot on 438.125 MHZ and the ZL3CHD repeater on 439.2125 MHZ.

E: DV	(Remain 378 CH)									
		Frequency						Call Sign		
CH	CH Select	Operating Freq	DUP	Offset Freq	TS	Mode	Name	Your	RPT1	RPT2
0	99	438.125000	-DUP	5.000000	25k	DV	Use Reflector	CQCQCQ	ZL3DW B	ZL3DW G
1	100	438.125000	-DUP	5.000000	25k	DV	Unlink	U	ZL3DW B	ZL3DW G
2	101	438.125000	-DUP	5.000000	25k	DV	Info	I	ZL3DW B	ZL3DW G
3	102	438.125000	-DUP	5.000000	25k	DV	Echo	E	ZL3DW B	ZL3DW G
4	103	438.125000	-DUP	5.000000	25k	DV	REF001 C	REF001CL	ZL3DW B	ZL3DW G
5	104	438.125000	-DUP	5.000000	25k	DV	REF030 C	REF030CL	ZL3DW B	ZL3DW G
6	105	438.125000	-DUP	5.000000	25k	DV	Kiwi XLX299 K	XLX299KL	ZL3DW B	ZL3DW G
7	106	438.125000	-DUP	5.000000	25k	DV	Tech XKX299 A	XLX299AL	ZL3DW B	ZL3DW G
8	107	438.125000	-DUP	5.000000	25k	DV	BM539 XLX299 B	XLX299BL	ZL3DW B	ZL3DW G
9	108	438.125000	-DUP	5.000000	25k	DV	QuadNet Tech	XRF757CL	ZL3DW B	ZL3DW G
10	109	438.125000	-DUP	5.000000	25k	DV	QuadNet Array	XRF757AL	ZL3DW B	ZL3DW G
11	110	439.312500	-DUP	5.000000	25k	DV	Use Reflector	CQCQCQ	ZL3CHD B	ZL3CHD G
12	111	439.212500	-DUP	5.000000	25k	DV	Unlink	U	ZL3CHD B	ZL3CHD G
13	112	439.212500	-DUP	5.000000	25k	DV	Info	I	ZL3CHD B	ZL3CHD G
14	113	439.212500	-DUP	5.000000	25k	DV	Echo	E	ZL3CHD B	ZL3CHD G
15	114	439.212500	-DUP	6.000000	25k	DV	REF001 C	REF001CL	ZL3CHD B	ZL3CHD G
16	115	439.212500	-DUP	5.000000	25k	DV	REF030 C	REF030CL	ZL3CHD B	ZL3CHD G
17	116	439.212500	-DUP	5.000000	25k	DV	Kiwi XLX299 K	XLX299KL	ZL3CHD B	ZL3CHD G

Figure 8: DV channels stored in the standard memory bank

Note that the reflector commands all have the module letter and an L to tell the repeater that you want a link. The I, U and E codes have seven leading spaces. RPT1 is the repeater node (R1), and RPT2 is the gateway node (R2).

Once you have entered a block you can copy it down and change the RF frequencies and the repeater fields to suit a different repeater.

Procedure

Start by adding a new bank name for your DV channels. **Memory CH > Bank Name.**

Select the new bank and enter the new channels there. They will automatically be added to the main channel list.

Entering a frequency without putting a number in the CH Select column, adds a new channel to the bottom of the existing channel list.

Over to the far right in the area marked Call Sign. You will see that the 'Your' field has defaulted to 'Use Reflector CQCQCQ.'

Double-click the **Your** field and

1. If you need to add a link to a reflector. Click the ... icon below 'Call Setting.' Erase the text and type in the reflector, module and L, e.g. REF030CL.
2. Click the ... icon below 'Access Repeater' and select **your repeater or hotspot**.
3. Repeater select should have changed to Gateway. If not, click the ... icon below 'Repeater Select' and select **Gateway**.
4. Click OK. This will automatically enter the correct data into the UR (Your), RPT1 (R1), and RPT2 (R2) fields.

Clone Write the revised configuration back to the radio and hit **Save** to hold a copy on your PC.

THE 4TH FIELD

The 4th field is the MY call sign field. It only has to be entered once. **Menu** > Set > **My Station** > **My Call Sign** > **select a slot** > **Quick** > **Edit**. Use the **Dial** and **CS** to enter your callsign and suffix, e.g. ZL3DW /ID52. **V/MHz** is a backspace. While you are in My Station, you can enter the TX message. Something like "*Andrew Christchurch.*"

USING DV MEMORY CHANNELS IN THE DV MODE

You can select the DV channels in the same way that you would select an FM channel. Rotate the **Dial** or **Multi** knob to select a reflector to link to and transmit briefly to activate the link. As usual, use the Use Reflector CQCQCQ channel to listen or talk on the reflector. Just make sure you have not inadvertently changed to a channel on a different repeater. You can use the unlink, information, and echo features the same as you would in the DR mode.

Once you have all the channels programmed, I guess this way works as well as using the 'Your Call Sign' list in DR mode. It is just clunkier to change repeaters.

NOT ON THE LIST?

What happens when you want to select a reflector that's not on your channel list? You can just set the fields manually.

You cannot set the D-Star repeater fields on a VFO channel. It must be a memory channel set to the DV mode. The easiest thing is to have a memory channel with a link to a reflector that you can change to a different reflector when you want to.

Manual field entry ID-51, ID-52

Select Menu > Call Sign. Select the UR line. Click Quick > Edit. Enter a reflector (or a gateway repeater link). For example, XLX299KL. Press the Enter (centre) button.

Use the same process to edit R1 and R2. R1 is the repeater or hotspot callsign and module. Make sure its module letter is in the 8th position. R2 is the repeater or hotspot gateway callsign and G for 'gateway.' Make sure the G is in the 8th position.

If you want to change the MY: field. Select the MY line. Click Quick > Edit then select an alternate callsign. Click Quick > Edit, on that line to edit or create a new MY entry.

Manual field entry IC-705, IC-9700, IC-905 etc.

Select Menu > CS. Touch and hold the UR: line. Touch Edit. Use the keyboard to enter a reflector (or a gateway repeater link). For example, XLX299KL. Touch ENT to save it. Touch Exit.

Use the same process to edit R1 and R2. R1 is the repeater or hotspot callsign and module. Make sure the module letter is in the 8th position. R2 is the repeater or hotspot gateway callsign and G for 'gateway.' Make sure the G is in the 8th position.

If you want to change the MY: field. Touch and hold the line then select an alternate callsign. Or touch and hold the line and select Edit, then create a new MY entry.

ONE-OFF NEW REPEATER

As a one-off, you can set the radio frequency to the local D-Star or multi-mode digital voice repeater, or your hotspot. The radio should automatically set the correct repeater offset, but you should check that it is correct. Set the DV mode and save the channel in one of the standard memory slots. Then you can set the D-Star repeater fields manually using the manual field entry just discussed.

Terminal modes

The Icom radios can be made to operate in two modes that interface directly with the Internet via your home LAN. I have no experience with the Icom Terminal modes, but online sources say that you can only make routed calls to specific callsigns or repeaters with it. **"The new Terminal and Access Point modes do not support linking to REF or any other reflectors. They only support Callsign Routing."** Since most users do not know how to route a call back to you, and most repeaters do not support routed connections anyway, this mode is of little use.

To establish a Terminal Mode, you have to open port forwarding of port 40000 on your home router to enable communication from the linked gateway. You also need to know the IP address of a G3 D-Star gateway that is using the RS-RP3C program. For more information, you can download a 29-page manual from Icom called, 'About the DV Gateway function.'

TIP: I believe that it is much easier to abandon the Terminal Modes and use an MMDVM hotspot instead. Hotspots are easier to use, and you do not have to open a port on your router to the outside world.

The Terminal Mode

The Terminal Mode lets you configure the radio to access gateway repeaters and reflectors directly over the Internet, without using a repeater or any radio transmission at all. Your radio is connected to your PC or an Android phone via a USB cable. You talk into your radio, but it does not transmit. Instead, the speech is carried over the USB cable to the computer or phone and from there, directly to a D-Star network. You can only make calls over the connected network node (repeater). You cannot extend that by making reflector or gateway links. This constraint makes the terminal mode much less useable than connecting via a dongle or a hotspot.

The Access Point mode

The Access Point mode is like the terminal mode except that a D-Star radio acts as a hotspot. Instead of talking into the connected radio, you talk into a handheld or mobile radio on a simplex frequency. The call is received by the connected radio and passed over the USB cable to the computer or phone and from there over the Internet to the D-Star node. Audio coming from the node is transmitted by the connected radio and received by the handheld (or mobile) radio.

This is a **very** expensive way to implement a D-Star hotspot, but I guess it could be useful if you want to make D-Star calls from around your yard. It has the same problem that you can only make calls over the connected node, not the entire D-Star network.

Terminal mode to a Pi-Star hotspot.

You can use Terminal mode to connect an IC-9700, ID-51 and other Icom radios, to a Pi-Star hotspot. See the video at https://www.youtube.com/watch?v=Iv8K4jgYMyU You need an OPC-2350LU cable to connect the Data jack on the radio, (not the USB port), to a USB port on the Raspberry Pi. Change the Pi-Star modem setting to **Icom Radio in terminal Mode (D-Star Repeater only)**. You also have to change the Controller Software from MMDVMHost to **D-Star Repeater**. The advantage of using this mode is that the radio does not transmit. I have tried this, and it does work. But I am just as happy with normal hotspot operation, where the radio does transmit. I set my radio for the lowest power output when using a hotspot.

Terminal mode software and cables – handheld radios

The ID-31A+, ID-31E+, ID-51A+2, and ID-51E+2 need an OPC-2350LU data cable and possibly a USB OTG (on-the-go) adaptor. The ID-52A and ID-52E use a standard micro USB cable available from any electronics or computer shop, (usually USB-C to micro USB).

Use the RS-MS3W software for connection to a Windows PC. Use the RS-MS3A software for connection to an Android phone. The phone can connect to the Internet via WiFi or cellular. Note that cellular data might be expensive.

You have to add the IP address or hostname of a G3 D-Star gateway that is using the RS-RP3C program into the RS-MS3A or RS-MS3W program or the connection will not work.

Terminal mode software and cables – desktop radios

The **IC-705** has WiFi so it can connect to your network without a cable or a physical connection to your PC. You can connect an **IC-9700** directly to your home network via the Ethernet port on the radio, or via an OPC-2350LU data cable to a USB port on your computer or an Android phone. For an Ethernet connection use **Menu > 2 > DV GW > Internal Gateway Settings > Terminal/AP call sign**, to enter your callsign and module letter.

The USB connection needs the RS-MS3W software for connection to a Windows PC, or the RS-MS3A software for connection to an Android phone. Enter your 'Terminal/AP' callsign into the software. You have to add the IP address or hostname of a G3 D-Star gateway that is using the RS-RP3C program into **Menu > 2 > DV GW > Internal Gateway Settings > Terminal/AP call sign, Gateway Repeater (Server IP Domain).**

The **ID-4100 A/E** is also able to use the Terminal and Access point modes. The connection method is similar. A USB connection to an Android phone or PC equipped with the RS-MS3A or RS-MS3W program.

Phone and PC applications

Icom has released several phone apps for controlling the radio via a phone and for transmitting pictures over D-Star. You can take a photograph, clip it to size in the app and send it to the D-Star radio for transmission to another D-Star radio. The 'picture sharing mode' is covered in more detail in the next chapter.

There is a handy 'Icom Application Guide' on the Icom website. The way that these applications connect to the radio depends on the radio model and not all models support all of the functions.

RS-MS1A

RS-MS1A is an Android phone app that lets you control the DR mode linking and unlinking of reflector or gateway calls from your phone. It works with all Icom D-Star radios. In most cases, you will need a data cable to connect the phone to the radio. Some radios are provided with one. The ID-52, IC-705, ID-4100, and ID-5100 can communicate over Bluetooth instead of using an annoying data cable.

You can take pictures on the phone and share them over the D-Star 'DV Fast Data' mode, which is much faster than the normal DV voice mode. Or you can elect to send the picture more slowly over a series of 'overs' while chatting.

You can use the app to send text messages over D-Star, much like a phone.

You can edit callsigns, use the 'Receive History,' and see the repeater list on the phone. The software also provides a very fast way to update the DV repeater list from the Internet.

Received D-PRS position data can be plotted on a map on the phone. This can be used offline, showing current and previously cached positions.

The fast data mode, transceiver control, and map functions are not available on the ID-31 or IC-7100. Bluetooth connection is only available on the IC-705, ID-52, and ID-4100 (with UT137 option).

RS-MS1I

RS-MS1I is an Apple iOS phone or iPad app that lets you control the DR mode linking and unlinking of reflector or gateway calls from your phone. It can also be used to share photos, plot positions on a map, and send text messages. The program is functionally identical to the RS-MS1A application.

RS-MS1I is only compatible with the IC-705, ID-52, and ID-4100 radios (with the UT137 option), because it always uses a Bluetooth connection.

Software	Connection	Radio models
RS-MS1A	Data cable or Bluetooth	All radios
RS-MS1I	Bluetooth	IC-705, ID-52, ID-4100
RS-MS3A	Data cable	ID-31+, ID51+2, ID-52, IC-705, ID-4100, IC-9700
ST-4001A	Bluetooth, WiFi->WiFi, or Ethernet->WiFi	ID-52, IC-705, IC-9700
ST-4001I	Bluetooth, WiFi->WiFi, or Ethernet->WiFi	ID-52, IC-705, IC-9700

RS-MS3A & RS-MS3W

RS-MS3A is an Android application for using the 'DV Gateway' or Terminal Mode function. It requires Android 5.0 or later and the OPC-2350LU data cable or a standard USB cable on the ID-52 and IC-705. The package is compatible with ID-4100A, ID-51A +2, ID-52, IC-9700 and IC705 radios. See the chapter on terminal Modes. RS-MS3W is the Windows equivalent of the RS-MS3A phone app.

ST-4001A & ST-4001I

ST-4001A is the Android version of the photo management software. ST-4001I is the Apple iOS version for iPhones or iPads. Both versions are compatible with the IC-705, IC-9700, and ID-52A. The program can send an image to the SD card in the radio. It could be a picture taken on the phone, received in an email or online chat, or downloaded from the Internet. It sets the image as the 'TX picture' on the SD card. Once in the radio, it can be sent using the D-Star 'DV Fast Data' mode, or more slowly over a series of 'overs' while chatting in the DV mode. You can add a text message such as your callsign to the image. See the 'Picture sharing mode' in the next chapter.

ST-4001W

ST-4001W is the Windows version of the photo management software. It is compatible with the IC-705, IC-9700, and ID-52A. The program can resize an image downloaded from the Internet or a camera etc. and place it onto the radio's SD card. Then it can be sent using the D-Star 'DV Fast Data' mode, or more slowly over a series of 'overs' while chatting in the DV mode. You can add a text message such as your callsign to the image. See the 'Picture sharing mode' in the next chapter.

CONNECTIONS AND APPLICATIONS

Radio models	Connection	Software
IC-705	Bluetooth	RS-MS1A, RS-MS1I
IC-705	OPC-2417 micro USB cable or OPC-2418 Type C USB	RS-MS1A, RS-MS3A
IC-705	WiFi to router, or WiFi to phone	ST-4001A, ST4001I
IC-7100	OPC-2350LU data to USB cable	RS-MS1A
IC-9700	OPC-2350LU data to USB cable	RS-MS1A, RS-MS3A
IC-9700	Ethernet cable to router. WiFi to phone	ST-4001A, ST4001I
ID-31 A/E	OPC-2350LU data to USB cable	RS-MS1A
ID-31 A/E+	OPC-2350LU data to USB cable	RS-MS1A, RS-MS3A
ID-4100 A/E	UT-137 option for Bluetooth	RS-MS1A, RS-MS1I
ID-4100 A/E	OPC-2350LU data to USB cable	RS-MS3A
ID-51 A/E +	OPC-2350LU data to USB cable	RS-MS1A
ID-51 A/E +2	OPC-2350LU data to USB cable	RS-MS1A, RS-MS3A
ID-51 A/E 50th An	OPC-2350LU data to USB cable	RS-MS1A
ID-5100 A/E	UT-133A option for Bluetooth	RS-MS1A
ID-52 A/E	Bluetooth	RS-MS1A, RS-MS1I ST-4001A, ST4001I
ID-52 A/E	OPC-2350LU data to USB cable	RS-MS1A, RS-MS3A

ST-4002A

ST-4002A is an Android application that inputs your position to the radio from the GPS receiver in the Android phone. It is useful for models that don't have a built-in GPS receiver. It works with the IC-7100, IC-9100, IC-9700, ID-4100, and ID-5100.

ST-4003A & ST-4003W

ST-4003A is an Android app that lets you set the time on your radio from the accurate time used by the cellular network. It works with the IC-705, IC-7100, and IC-9700. ST-4003W is the Windows equivalent.

The IC-9700 can also set the time from an NTP Time Server, via the Internet if an Ethernet connection to your home LAN is connected. Or it can sync the time to GPS if an external GPS receiver is connected.

The IC-705 can set the time from an NTP Time Server via the Internet if a WiFi connection to your home LAN is enabled. However, it is easier to sync the time using the internal GPS receiver. Set **GPS Time Correct** to **Auto** and the time will be automatically set when the radio receives a GPS signal.

RS-BA1 V2

RS-BA1 V2 is Icom radio remote control software, primarily for the Icom HF radios, although it is compatible with the IC-9700 and the IC-705.

DOOZY

Doozy is a Windows program. Written and developed by David Grootendorst PA7LIM. See https://www.pa7lim.nl/doozy/. It is an easier way of using Icom radios in the 'Terminal Mode' where they don't transmit. They just communicate with the network via your PC. There is Icom Gateway software to do this, but it is difficult to set up and you have to open port forwarding on your home network router. Download Doozy at http://software.pa7lim.nl/Doozy/.

Doozy will work with the Icom ID-52 and IC-705 using a standard USB cable. It also works with the IC-9700, ID-51+2, and ID-31 if you have the OPC-2350LU data cable. Also, the ID-4100 (probably), but this has not been tested.

IRCDDB REMOTE

ircDDB remote is an Android or Apple iOS phone app also written and developed by David Grootendorst PA7LIM. It talks to your Pi-Star hotspot over the WiFi connection. I find it great because you can save several reflectors and choose between them.

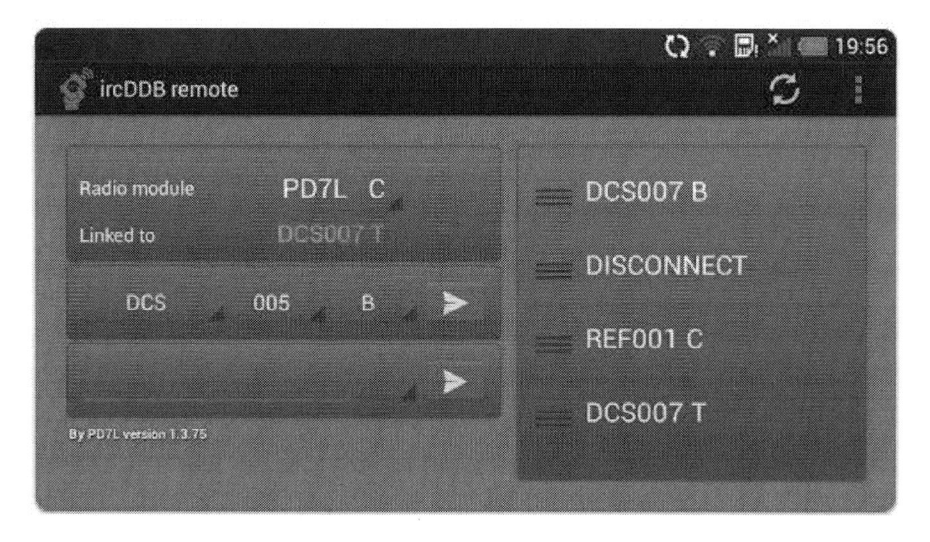

Figure 9: ircDDBremote interface, iOS and Android

The app does not have an option for linking to XLX reflectors, you can choose from DCS, REF, or XRF. However, if you enter an XLX reflector number into the search box, (or the box below the Refresh button in the iPhone version), it will add it to your favourites list, and then you can link to it.

PEANUT

Peanut is yet another PA7LIM creation. It is a phone app for Android phones that supports connection to various D-Star reflectors and DMR talk groups. There is also a Windows PC version.

The Peanut project provides access to some reflectors but not all. You don't need a dongle or a digital voice radio. The phone or PC does the routing to the network.

You do have to be D-Star registered to access the DPlus (D-Star) reflectors and registered for DMR if you want to access DMR talk groups. To do that you must have an amateur radio callsign. You also need a 'Peanut' code from https://register.peanut.network/. Some of the reflectors are transcoded so that you can talk between digital voice modes.

The Peanut dashboards are at,

- Global http://peanut.pa7lim.nl
- Japan http://peanut.xreflector-jp.org
- USA http://peanut-usa.pa7lim.nl

BLUEDV

BlueDV by PA7LIM is available in versions for Linux, Windows, Android, or Apple iOS. It is another application that can be used without a radio, although it does need a D-Star dongle. It also works with some hotspots which will need a serial connection to your PC or a Bluetooth connection to your phone. You can listen to three digital voice modes at the same time

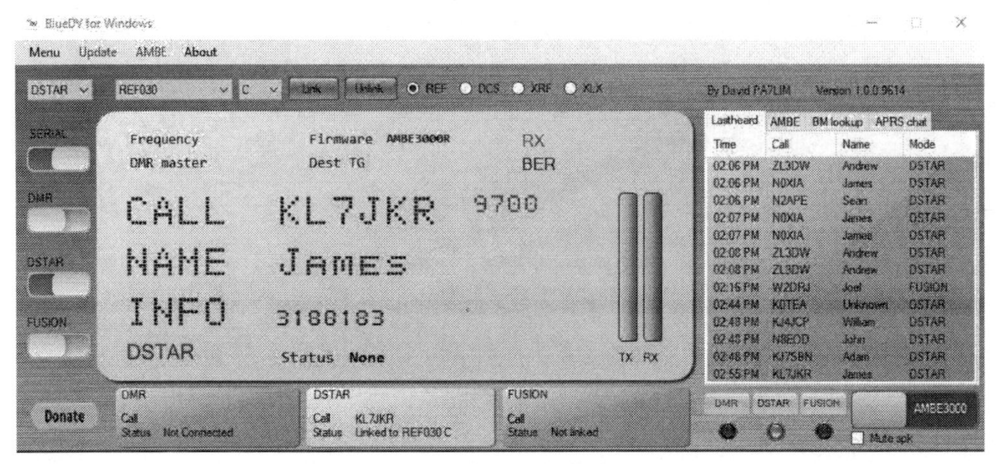

Figure 10: Blue DV by David PA7LIM (Windows version)

BlueDV will work with,

- Portable AMBE Server https://reflectorloversclub.jimdofree.com/
- the NWDR ThumbDV, (currently out of stock) https://nwdigitalradio.com/products/thumbdv%E2%84%A2)
- the DVMEGA DVStick 30 and Globetrotter, see https://www.dvmega.nl/
- the BlueStack hotspot https://www.combitronics.nl/index.php?route=product/category&path=62 and the
- ZUMspot AMBE Server, see https://www.hamradio.com/detail.cfm?pid=H0-017021

ZUMspot AMBE Server

The ZUMspot AMBE Server has an AMBE3000 chip on board. It can be used with BlueDV, Peanut, DummyRepeater, or Buster software.

http://www.pa7lim.nl/bluedv-windows/ http://www.pa7lim.nl/peanut/
https://github.com/g4klx/DummyRepeater
https://apps.apple.com/us/app/buster/id1060175273?mt=12

D-STAR DONGLES

I have experimented with two D-Star dongles. They allow you to access the DV modes without a radio.

ThumbDV AMBE 3000 USB stick with BlueDV

The Thumb DV by www.NWdigitalRadio.com looks similar to the DVMEGA DVStick 30, but the board layout and chipset are different. I purchased a pre-owned one from TradeMe which is our local online auction site.

The dongle will work with a PC running BlueDV, or an Android phone running the BlueDV AMBE (Beta) software. I had some difficulty getting started. The software hardly ever recognised the dongle.

I reloaded the USB driver from https://ftdichip.com/drivers/vcp-drivers/ and at the same time, I added a YSF repeater.

The BlueDV website recommends this if you have problems. I don't know which action fixed the problem. But after doing both, the software works fine.

If the top line of the main BlueDV display says **Firmware AMBE3000R** and you can operate the DMR, D-Star, and/or Fusion switches, the connection is working.

TIP: If the receive audio is distorted, update the USB driver. You must use the **460800** *baud rate (except for very early ThumbDV dongles). If you have connection problems, and you have already updated the USB driver, look up the Com port in Windows Device Manager and set the Port Settings to* **460800, 8, None, 1, None**.

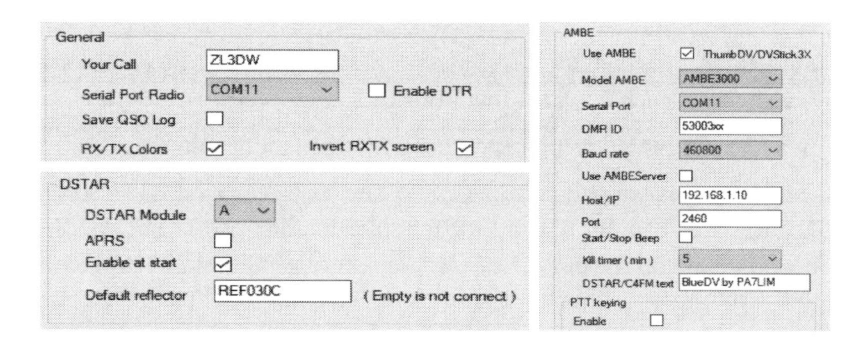

To use the Android phone option, download and install the BlueDV AMBE (Beta) software for the phone from the Google store. The dongle is plugged into your PC which has to be running the AMBE Server utility, https://www.pa7lim.nl/garage/. BlueDV AMBE is not available for the iPhone.

Portable AMBE Server

The Portable AMBE Server is another D-Star dongle that can get you onto D-Star reflectors without a D-Star radio. It is available from the XRF reflector lovers club in Japan. The unit contains a Raspberry Pi Zero W and a modem board containing the AMBE 3000 vocoder chip. It supports D-Star, C4M (YSF), and DMR. The unit can be powered by a USB power supply or a USB port on your PC. The power connector is the micro USB port closest to the short end of the case and the LED.

There are three ways to use the AMBE Server dongle. You can connect the unit to your home WiFi and use it from within your home network with BlueDV running on a computer or BlueDV AMBE (Beta) running on an Android phone. BlueDV AMBE is not available for the iPhone. Or you can use the 'tethered mode,' which involves configuring your Android phone to act as a WiFi hotspot and connecting the AMBE Server dongle to it. The third option is to plug the unit into a USB port on your PC and run the AMBE Server utility, available at https://www.pa7lim.nl/garage/. In this scenario you can access the AMBE server from an Android phone anywhere in the world. The phone connects to the AMBE server at home via the cell phone system or any connected WiFi such as a hotel or restaurant free WiFi service. You will be able to make D-Star, YSF, or DMR calls from anywhere. I have not tried the tethered or remote options because I don't have an Android phone. There are setup videos for the normal WiFi and tethered modes on the website.

Setting up the Portable AMBE Server for local WiFi was quite straightforward.

1. Remove the four black case screws only
2. Remove the SD card from the Raspberry Pi. It just slides out – no latch.
3. Insert the SD card into a card reader and plug that into a USB port on your computer.
4. Very important! Ignore all the Windows error messages and do NOT format the SD card. Close each message that pops up.
5. Windows Explorer should pop up with the SD card 'boot' drive directory
6. Run AMBEDconfig.exe. Change the SSID and password message to your WiFi ID and password. Input the AMBE Server IP address 192.168.xx.131/24. It is labelled wlan0 on the box. The first three numbers should be the same as the WiFi router's IP address (often 192.168.1). I had to change the third number from 43 to 1.
7. WiFi router's IP address, e.g. 192.168.1.8. It should be in your phone's WiFI setup information. Click **Save**.
8. Remove the SD card and replace into the Raspberry Pi. Power up the AMBE Server dongle. Open BlueDV setup. Check Use AMBE and Use AMBE Server. Set Hots IP to the AMBE Server IP address (192.168.1.131). **Save > Serial**.

D-Star Picture Sharing mode

The picture-sharing mode was introduced in a firmware update for the IC-9700 and has since been included in the IC-705 and the ID-52 as well. It allows you to send and receive photos over D-Star.

Not wowed so far? The neat feature is that with the free Icom ST4001 software for Android, IOS, or Windows you can take a photo with your phone, copy it to your radio via a Bluetooth, WiFi-router-WiFi, or Ethernet-router-WiFi connection, and transmit it to another station. You can even add a caption such as, "Here is a photo of us at the beach!" or "This is my latest toy, an Icom IC-9700!"

It seems complicated initially, but it is actually quite easy. The function works very well either directly on the radio or using the free Icom software. I am not sure why sending pictures is not more popular. Perhaps it's because only three models support the picture-sharing mode.

Please note that sending pictures over D-Star is not private. Avoid sending photos of a sensitive nature. Any pictures transmitted over D-Star will be received by every compatible D-Star radio using the same repeater or reflector.

PICTURE SHARING ON THE IC-705 AND IC-9700

Icom has a help file for the 'Picture Sharing Mode' for the IC-705 and IC-9700 at, https://www.icomjapan.com/support/manual/2159/

Picture menu

MENU > 2 > Picture, opens the 'Picture Sharing' mode. You can view pictures received over D-Star and send pictures that have been stored on the SD card.

While you can take the SD card out of the radio and load the images onto the card from your PC or some cameras. There is a better and much cooler way! Using the free Icom ST-4001I or ST-4001A software, you can download an image from your phone directly to the radio. This is fantastic for holiday or portable operation because you can take photos on your phone and immediately send them out to your friends over D-Star.

RX

Touch the RX image to get a larger image of a received photo. You also get information about who sent it, the image size, and the quality setting that was used. Select the HISTORY Soft Key to get a full-screen image or step through previously saved images. You can also save the picture as a .jpg file. When you receive a transmission that includes a picture the RX Picture icon will be displayed on the main display. You can turn it off, but why would you? If you happen to have the Picture Sharing mode onscreen when the radio is receiving an image, you will see the picture appear block by block. But the image will still be received even if your radio is not in the Picture Sharing mode.

Depending on the transmission method and the picture size and quality, it may take several overs before a complete picture is received.

TX

The TX image is shown as a group of blocks. Each block is sent as a data burst. If you select a high resolution, there are more blocks, so the image takes longer to send. This does not matter for simplex contacts, but it could annoy D-Star repeater or reflector users. The top left square should be red indicating that it is the first block to be transmitted. The red block moves across and down the picture as the image is being sent. To select a different image, touch the TX image, then touch it again or TX SET, then touch the TX Picture line and use the up and down arrows to select a different picture. Finally, touch the SET Soft Key to select it. Then touch Back ⮌, Back ⮌.

TX SET lets you change the picture size. The default size is 320x240. You can also send at three different quality levels. I suspect this represents three levels of data compression.

RECEIVER. You can change the data that would normally go in the TO box (UR data field). CQCQCQ will send the picture to anyone who has a radio capable of receiving it. Or you can enter a destination callsign. Note that this does not make it private. It only lets people know who you wanted to send it to.

HISTORY shows you images you have transmitted before and allows you to re-transmit them.

BLOCK lets you change the start position in case you want to leave out the top part of the image.

1ST. If the red block is not top left and you want to start again. Touch and hold the '1ST' Soft Key.

ⓘ Indicates the IP address and network name (if set). This is only relevant if you are using the Ethernet connection to upload photos to the radio.

PICT TX

If you select PICT TX, (while in D-Star mode), an icon will appear at the top of the display indicating that a picture will be transmitted with your next D-Star transmission. The picture will be sent along with your voice transmission and will appear on the Picture RX screen at the destination radio.

When you stop transmitting, the picture will pause. On your next few overs, the picture transmission will continue from where it stopped on the previous transmission.

This is the "slow data mode." It may take several overs before a full image, particularly a large or high-quality image, is completed.

TX ALL

The TX ALL mode sends the whole picture as a high-speed data transmission using the 'DV Fast Data' mode. It is much faster than the PICT TX mode, but you cannot talk at the same time. Announce that you are going to send a picture, then send it using TX ALL.

TIP: This is definitely the best way to send a picture. It gets the job done in one transmission.

PICTURE SHARING ON THE ID-52

You can use the free Icom ST4001A or RS-MS1A software for Android, or the ST4001I or RS-MS1I versions for Apple iOS to transfer a photo from your phone or PC to the radio via Bluetooth. Once the picture is on the radio's SD card it can be transmitted via D-Star. You could also remove the SD card from the radio and copy pictures from your PC onto it. To view the pictures on the SD card use; MENU > SET > SD Card > TX/RX Picture View.

The picture-sharing screen

MENU > PICTURE shows the picture-sharing screen below the frequency display. The left side shows the most recent received picture (or blank). The right side shows the image that will be transmitted. A small red block should be in the top left corner it shows the first data block to be transmitted. As the image is sent the red block scrolls across and down the image. To close the screen, press QUICK and select Exit picture screen.

TIP: In the unlikely event that you are receiving images on the main band and the sub-band at the same time, both images will be stored, but only the picture being received on the main band will be displayed on the picture-sharing screen.

If you don't have the picture-sharing screen open, the RX Picture icon is shown on the main display while the radio is receiving and after it has received an image. This function can be turned off. I don't know why you would want to do that.

Received pictures on the ID-52

Enter the picture-sharing screen using MENU > PICTURE use the left or right pad to select the received picture on the left side and press the Enter (centre dot button) to open a larger view of the image.

Use the up and down pad buttons to view previously received images.

Press the Enter (centre dot button) to open the picture detail screen. It shows the image number, its size, the quality mode selected to send it, the callsign of the sender, and the date & time it was received. The up and down pad buttons can be used to view previously received images.

Sending pictures on the ID-52

Enter the picture-sharing screen using MENU > PICTURE. If the picture you want to send is displayed, you can send it immediately. Press QUICK > TX All to send the picture immediately in the DV Fast Data mode. This is almost always the best way to send a picture to someone. But if you want to chat, you can send it the slow way while you are talking.

To set up the image to be sent while you talk, enter the picture-sharing screen using MENU > PICTURE. If the picture you want to send is displayed, you can send it on your next 'over.' Press QUICK > Picture TX > Single TX. If the image transmission has not been completed when you stop talking, it will carry on when you begin the next over and continue over the next few 'overs' until it has all been sent.

NOTE: If you press the PTT or press the PTT and talk while an image is being sent in the fast DV mode, the radio will switch to the slow normal DV method and continue to send the image at the slower rate while you talk.

NOTE: Not all D-Star radios have the fast data mode. They will only be able to receive images sent in the normal DV mode. Also, some radios don't have a screen to display an image on.

Changing the picture that you want to send (ID-52)

Enter the picture-sharing screen using MENU > PICTURE use the left or right pad to select the transmit picture on the right side and press the Enter (centre dot button) to open a larger view of the image. Press Enter again and select TX Picture. Use the Dial to select an image to transmit and press Enter to select it.

Optionally, select a picture size. Setting a smaller image will make the transmission faster. You cannot select a size that is bigger than the size of the image on the SD card.

Optionally, select a picture quality. A lower-quality image will be sent faster than a high-quality image. The default is 50%. You can set the callsign of the intended Receiver.

Note that sending pictures is not private. Anyone that receives the signal from your local repeater, and anyone on the reflector or connected gateway repeater, will receive the image. The **Receiver** *setting is just an indication of who the image was for.*

You can move the TX block marker if you only want to send only the lower part of an image. It is not a very useful function, but it's there.

After you have selected the image that you want to send, you can send it using either the fast DV data mode or the slower normal DV mode, as described in the previous section.

Figure 11: The ST4001I phone app

RULES FOR PICTURES

Pictures must be in JPEG .jpg format. They can be 640x480 pixels, 320x240 pixels, or 160x120 pixels. Icom says that other sizes cannot be read by the transceiver.

Smaller pictures are faster to send. 640x480 or 320x240 files can be sent as smaller images by changing a menu setting in the radio. The file size must be less than 200 kB. The file name can be up to 23 characters plus the three-character .jpg extension.

The radio can only display 500 pictures even if more are held on the SD card.

PICTURE SHARING ON THE ID-51 +2

The ID-51 does not support the picture-sharing mode, but you can send a picture to a radio that does, using the Icom RS-MS1A software for Android connected via an OPC-2350LU data cable. The radio does not know that it is a picture that you are sending, it is just sending data from the phone.

Icom Radio SD card

The newer D-Star radios have an SD or micro SD card slot. The SD card holds different things on different models. Items can include,

- Radio configuration settings
- Memory channels
- FM radio memory
- GPS memory
- GPS location log (ID-52, IC-705)
- Communications log
- Recorded signals off the air
- Automatic answering audio
- RTTY decode logs (IC-705, IC-9700)
- Voice recorder (ID-52)
- Voice message memories
- D-Star received pictures and pictures to be transmitted
- UR callsign memory (I use it for reflectors)
- Saved screen capture images
- Uploading new firmware
- Start-up screen image
- D-Star repeater list

Unfortunately, the SD card is not supplied with the radio. You have to buy one. The IC-9700 requires a full-sized SD card. The ID-51, ID-52 and IC-705 use a micro SD card. The transceiver creates a directory for the radio model with up to 12 subdirectories. You can use the same card for more than one radio model.

You do not need a very big or very fast SD card. Don't pay extra for a super-fast SD card. Buy the smallest capacity you can get. You can use a 2 Gb Micro SD card or an SDHC card from 4 Gb up to 32 Gb. Icom recommends using SanDisk© SD cards. But any SDHC card except Adata should be fine. I am using a 4 Gb SanDisk Micro SDHC but these days it is hard to find a card smaller than 32 Gb. I leave the SD card in the radio all the time unless I am transferring images to my PC or uploading new firmware to the radio.

Icom recommends formatting the SD card using the function in the radio, before using it for the first time. **MENU > SET > SD Card > Format.** I did not format my SD card and it works fine. Don't format the card if it already has files from another radio.

There is also a menu option to 'unmount' the SD card before you remove it. The same as you would when removing a USB stick from your PC. Use MENU > SD Card > Unmount. I don't bother using it and haven't had any problems, but it is your risk.

The blue SD indicator in the top right of the display just to the left of the clock indicates that there is an SD card in the SD card slot. A flashing SD icon indicates that the radio is writing information to, or reading from, the SD card.

SD card reader & writer

If you have a notebook PC, it may have a built-in SD card reader. If it does, it will probably be for the full-size SD cards, but you can buy a micro SD HC card that comes bundled with an adapter. If like me, you use a desktop PC without an SD card reader, you can buy a USB SD card reader. They write as well. They only cost $10 or £4 or thereabouts. You can use one that takes full-size SD cards and buy a micro SD card that is packaged with an adapter, or you can buy one that takes the micro SD card. I have one of each. USB card readers are available from electronics and computer shops, or the usual online sources.

Save the default settings ID-52, ID-51 +2, IC-705, IC-9700, and others

Icom recommends saving the default settings in case a problem affects the radio configuration. Menu > Set > SD Card > Save Setting. Select New File then press Enter, then Yes. If you need to recover the data use Menu > Set > SD Card > Load Setting. Select the latest file then All, or Except My Station (which excludes the Repeater List and MY callsign list), or Repeater List Only. The IC-705 and IC-9700 have options for All, Select, and Repeater List Only. You are also asked about keeping the SKIP settings on the repeater list. It is usually best to select Yes. You can import and export .csv files on the IC-705 and IC-9700.

Set screen capture to the Power button ID-52, IC-705, and IC-9700

You can set a function that will save a picture of the screen when you press the power button (short press). It can be annoying, so I recommend leaving it off until you want to save a copy of a screen. MENU > SET > Function > Screen Capture [PWR] Key. On the IC-9700 and IC-705 it is MENU > SET > Function > Screen Capture [POWER] Switch. Screen capture is not available on the ID-51.

Voice message TX ID-52, ID-51 +2, IC-705, IC-9700, and others

You can record a voice message on the ID-52 & ID-51, or 8 on the IC-705 and IC-9700.

Configuration Software

DOWNLOAD THE CS CONFIGURATION SOFTWARE

Although you can program any of the radios from the radio keypad, you will quickly find that this is tedious. If you are adding a lot of channels, it is certainly best to download and install the free Icom configuration software (CS).

Each Icom model has unique configuration software. As far as I know, it is only available for the Windows platform. I downloaded mine from the 'global' Icom website in Japan, but it should also be available on your local Icom website. Select **Products > Amateur > Support > Firmware/Software**. Then enter the model number of your radio into the search box and click **Search**. For example, IC-705 or ID-52A.

Then click the red link beside 'Programming Software.' If you are really keen, read the terms and conditions. Then click the checkbox and the **Download** button. Install the software as you would any other Windows program.

RT SYSTEMS CONFIGURATION SOFTWARE

RT Systems also makes programming software for the Icom radios, and they have a wide selection of interface cables for the radios that need one.

The Icom software has a similar basic capability to the RT Systems programmer. However, the RT Systems programmer has a better layout which makes accessing the data easier. You can cut-n-paste directly from an Excel file provided the data you paste in has the headers included and the correct data fields.

The software for the IC-9700 and most other Icom radios uses the 'Clone' mode to read and write data to and from the radio, so the transmitter does not come on during the upload or download process. A warning message about turning the radio off before attaching a cloning cable is displayed, but you can ignore it since the transfer is done over the standard USB connection between the radio and the PC which will already be connected. You do have to turn the radio off and then on again after sending an update to the radio.

The main screen on the IC-9700 version has nine tabs across the bottom. This is due to the multiple memory banks in the radio. There are 99 memories for each of the three bands, scan limits, memo pad, DR (D-Star) memories, call channels, satellite memories and GPS memories.

The top menu has a special D-Star menu option which lets you add D-Star reflectors to the main memory banks. This is a neat function because you can't easily save reflectors with the radio's menu commands. It automatically saves separate channels for the U, I, E, and CQCQCQ modes.

Make sure that you select the first memory slot number that you want to use for the new information. If you leave it at the default setting of '1' the existing data for four or more channels will be overwritten.

READING AND WRITING DATA TO THE RADIO

The Icom CS software uses the 'clone' mode to update the radio over a USB cable. Reading data from and writing data to the radio is very much the same as programming an FM handheld using PC software. To make changes, you start the CS (configuration software), connect the radio, upload the current configuration from the radio or a saved file, edit the configuration, save a backup on the PC, and then send the updated file back to the radio. It only takes a few minutes.

TIP if you have an old version of the CS software it may refuse to communicate with the radio. I had this happen and it was confusing. Download the latest version from the Icom website.

If you upload the updated configuration back to the radio and you get a message that says you are saving an old format, you can proceed with the upload. But it would be wise to update the radio's firmware and then upload the file again.

Programming cables

The ID-52A and the IC-705 use a standard 'micro USB' to 'USB Type A' cable. You can buy one at any computer or electronics store. It plugs into the same connector on the radio as the battery charger. The cable can be used to charge the radio, transfer GPS location data, CI-V rig control, and programming. If you don't have a cable, you can remove the SD card from the radio, download the radio config file from the CS software onto the SD card, and then reinstall the card in the radio. Using a USB cable is a lot easier.

The IC-9700 has a 'Type B' USB jack. It is the connector that you would see on a USB peripheral such as a printer. The 'USB type B' to 'USB type A' cable is used for CI-V control and the data connection when in the DD mode. The radio also has an Ethernet port.

The ID-51 and most other Icom radios require the OPC-2350LU data cable.

TIP: Be very careful not to accidentally hit the PTT (Push to Talk) button while you have the USB or data cable plugged into the radio, or while it is sitting in the charger. RF energy can be coupled into your computer or the computer part of the radio causing lockups and possibly failure of the radio. It is very easy to push the PTT while you are plugging in the programming cable. I try to get into the habit of turning the radio off while I plug it in. Or you could make sure to only plug in the computer end of the cable after you have plugged in the radio end.

You do not have to use a USB 3.0 port on the computer. The data transfer is not particularly fast, so a USB 2.0 port is fine. However, a USB 3.0 port is capable of supplying more current so it should charge the radio battery faster.

TIP: It is possible to access the files on the micro SD card in the ID-52A handheld from your Windows PC, over the USB cable. You could use this to upload backup files, new firmware, or pictures, and to download recordings and log files. Set **MENU > SET > USB Connect** *to* **SD Card Mode.**

Connect the cable

Plug the cable into the radio and any USB port on the PC and turn the radio on. Windows 10 will recognise the device. It is likely that you will have to download a device driver. You may have already done this for using the radio with digital mode, or logging software. If you do need a driver, it is available on the Icom website at https://www.icomjapan.com/support/manual/3193/ USB Driver IC-705/ID-52A/ID-52E/RS-BA1 Version 2. This driver supports the ID-52A and the IC-705. There is a different driver at https://www.icomjapan.com/support/firmware_driver/1974/ for the IC-7100, IC-7200, IC-7300, IC-7410, IC-7600, IC-7610, IC-7850, IC-7851, IC-9100, IC-9700, and IC-R8600.

The USB device driver for the IC-9700 and IC-705 will establish two virtual COM ports on your computer. Use the lower COM port number for CI-V and the CS software. It will also create an audio codec that is used for audio connections to PC software. The USB device driver for the ID-52A will establish one virtual COM port on your computer. I did not need to load a driver for the OPC-2350LU data cable and ID-51A, but it also creates one COM port.

Setting the COM port in the CS software

Click **COM Port > Setting** on the top menu of the CS screen. Enter the radio's COM port number from the dropdown list and **OK**.

If you don't know the radio's COM port number, click the **Search** button. It should find the radio. If Search does not find the radio, you can use Windows 'Device Manager' to find the COM port for your radio. The radio must be turned on. Look for a COM port that appears when you plug in the USB cable between the radio and the PC. If the radio creates two COM ports use the lower-numbered one.

USING THE CS-52 SOFTWARE

The CS-52 software is for the ID-52. You can use it to make changes to your radio configuration and to make a backup to use if there is a problem.

TIP: Always accept the option of creating a backup on the SD card before performing a firmware update. Some updates delete all your D-Star settings and saved memory entries.

Turn the radio on, connect the USB cable and start the CS software.

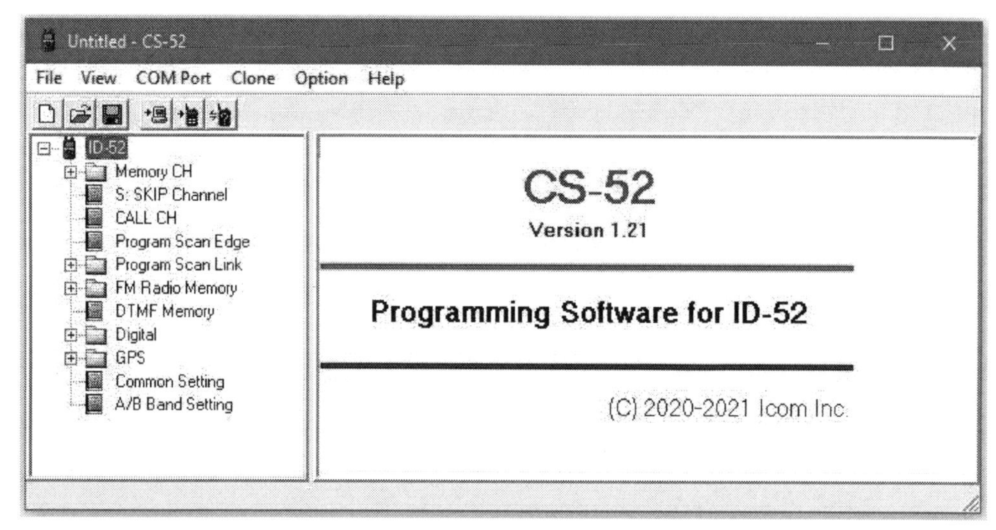

Figure 12: CS-52 main screen

Memory CH folder ID-52

Opening the memory channel folder reveals up to 100 groups. Each group can hold up to 100 stored memory channels, with a total memory capacity of 1000 'normal' channels. Channels can be in more than one group. For example, you might have a group with all New Zealand channels, another for Local channels, and another for 70cm repeaters. Some channels might be in all three groups.

Right-click a folder to import a group from a .csv formatted data file. It is best to enter one channel and use **right-click > export** to export a file to your computer. That will provide the column headers for the repeater file that you wish to import. You can import a .csv file that includes multiple groups.

TIP: This is 100% better than the Yaesu programming software for YSF radios which does not include column headers leaving you to guess what data is supposed to go in each column.

S: SKIP Channels ID-52

The radio can hold up to 100 Skip channels in 'group S.' You can enter any frequency not just memory channel frequencies. There is a choice of OFF, SKIP, or PSKIP. Frequencies marked SKIP are excluded from the memory scan. Frequencies marked PSKIP are excluded from the VFO scan and the memory scan.

TIP: You would normally save Skip channels during a scan. Not in the CS software. If the scan stops on noise or a held carrier, you can add the frequency to the SKIP list so that the scan won't stop there the next time around. A very cool and useful feature.

If you mark a channel as OFF, it will be scanned, but the frequency is in the list, so you can change the skip status easily in the future.

CALL CH ID-52

There are four 'Call Channels,' two on the 2m band and two on the 70cm band. They are a quick way to get to channels that you use a lot, such as your local repeater or a simplex frequency that you use all the time.

Program Scan Edge ID-52

You can set limits for up to 25 programmed scans. For example, you might create programs to scan the FM part of the 2m band, the whole 2m band, the Aircraft band, and the FM Marine band. The available modes are set according to the band of frequencies that you select. For example, you can set AM for Aircraft band reception.

Program Scan Link ID-52

This is almost too much flexibility. I doubt that very many people will ever get around to using this feature. I'm sure that I won't.

The idea is that you can create a scan list with gaps. The standard 'Program Scan Edges' scans all frequencies between upper and lower limits. The 'Program Scan Link' list, links two or more program scans together. Say for instance that you had a program scan that covered the FM repeater section of the 2m band and another that covered the FM repeater section of the 70cm band. You can create a Program Scan Link that extends the scan so that it covers the 2m repeaters and then the 70cm repeaters. i.e., all repeater outputs on both bands. I created one called "Bob." It could be useful if I took the radio away and did not know the local channels. However, since I have all NZ FM repeaters in the memory channels, I could just scan that.

FM Radio Memory ID-52

The FM Radio memory can hold up to 500 channels in 26 groups. You can import or export a list in .csv format.

DTMF Memory ID-52

The DTMF memory holds up to 16 DTMF short codes, used to control repeaters and link or unlink reflectors. You would probably normally program this on the radio rather than the CS software. Each DTMF channel can hold 24 characters. See DTMF codes on page 46. Higher numbers in 'DTMF Speed' send the codes more slowly.

T-Call is only used on E-version radios, for European repeaters, it transmits a 1750 Hz 'in band' tone burst.

One channel can be marked with S. The DTMF code on that channel will be sent if you press the SQL key while transmitting.

Digital ID-52

The Digital folder holds the settings and lists for D-Star operation. Many of these settings and stores would normally be loaded with data while you are using the radio, rather than editing them in the CS software. This is another good reason to always load the current configuration from the radio into the CS program, before making any changes. If you rely on a version saved on your PC, you may unexpectedly change settings.

Since I can't make the callsign routing mode work, I use the rather misnamed '**Your Call Sign**' list to store reflectors that I want to access often in the DR mode. You can assign a name to each entry. I saved CQCQCQ, E, I, and U as well so that I can use the Dial knob to link to a reflector and then change to the Use Reflector CQCQCQ mode. The list can hold 300 entries.

The **repeater list** may already be populated on the radio. It can hold the worldwide list of D-Star repeaters. This enables the 'Near Repeater' function which lets you select repeaters that are near your current location according to fixed data or the GPS location. The list is supposed to be used for making routed 'Gateway CQ' calls to a remote repeater. But that does not usually work. Most D-Star users will never use the gateway mode. The ID-52 can store 2500 repeaters.

My Station ID-52

The '**My Call Sign**' list can hold up to six callsigns for D-Star operation. It could be for six people that use the radio, or for six variants of your call sign. For example, ZL3DW / ID52, ZL3DW / SOTA, ZL3DW / MOB. The line marked with S for 'select' is the version that will be transmitted. You can change to a different version, on the radio at any time.

TX Message is a short message to be transmitted in the D-Star mode, along with your callsign. For example, "Portable at Lake Louise." You can store up to six messages. The one with the S will be selected. Or you can turn the message feature off.

TX History ID-52

The TX History stores the repeaters you have used most recently, who you called, and the reflectors you connected to. This screen is not intended to be edited with the CS software, but you could use it as a short-term memory bank.

RX History ID-52

The RX History shows the history of received calls. It is not intended to be edited with the CS software. 'Rx RPT1' is the gateway the call came through, usually, your local repeater and 'Rx RPT2' is the D-Star identity of the repeater. If you are using a hotspot, 'Rx RPT1' will probably be your call with a G module and 'Rx RPT2' will probably be your call with a band module (A, B, C).

Digital Setting ID-52

This page sets all the D-Star digital settings on the radio.

Tone Controls

- Tone controls for received audio

DV Set Mode

- Auto Reply sets an automatic voice message reply to a call made to your callsign. The 'Position' setting is a great safety feature. If someone calls you, the radio responds with your call sign and location. So, if you were injured on a trail or something like that, the radio would send your location. It could also be useful if you were mobile and could not talk on the radio immediately. Sadly, Auto Reply is unlikely to work because it relies on callsign routing.
- The DV Data TX function sets whether data is sent immediately after it is received from a connected device such as a PC or a weather station etc. Or if it has to wait until you press the PTT.
- DV Fast Data TX Delay (PTT), sets what happens if you press the PTT while the radio is sending fast data (usually a picture). The function only works if The DV Data TX function is set to PTT.

  If DV Fast Data TX Delay (PTT) is set to OFF, the radio switches to slow data while the PTT is pressed and then back to receive when you release the PTT. The remaining data will continue to be sent in the slow picture mode each time you transmit by pressing the PTT.

  If the item is set between 1 and 10 seconds, the radio switches to slow data while the PTT is pressed and then back to fast data for the nominated time when you release the PTT. If all the data has been sent, the radio returns to receiving. After the nominated period, the radio returns to receiving anyway.
- Digital Monitor can be set to listen to the channel in DV or FM, or automatically switch between DV and FM.
- Digital Repeater Set, selects whether you want the repeater callsign to be corrected if it is different to the setting you have in the FROM box. I can't see any point in turning this off.
- DV mode auto detect, switches the radio to DV if a digital signal is received and to FM if an FM signal is received. Yep, that sounds pretty useful! Leave it switched on.
- RX Record (RPT). The transceiver can record up to 50 calls. You can select whether to record all calls or only the most recent/current one.

- RX->CS key. If this is set to Call Sign Capture, holding down the RX->CS key will store the current, or most recent, callsign into the RX History. If you rotate the dial while holding down RX->CS, you can select a station to call. If this is set to RX->CS list, Rotate the dial while holding down RX->CS, to select a station to call. I don't think the callsign routing will work anyway.

Display

- RX Call Sign Display selects what happens when you receive a signal. In the normal mode, the calling station's callsign and message are displayed once. In the RX Hold mode, the callsign and message are displayed and then the callsign remains displayed. I find this very useful as I always forget the calling station's callsign. The Hold mode is the same as the RX Hold mode except the message and callsign continue to scroll every two seconds until another call is received.

- The RX Position Indicator setting controls whether the radio will display received position information. The RX Position Display Timer sets how long the position will be displayed.

- Reply Position Display sets whether a signal sent back via the auto-reply mode will show the other station's position. As this is pretty much the main reason to use the auto-reply feature, I would expect that you would leave this turned on.

- RX picture indicator sets whether the received picture icon will be displayed when a picture is received. I don't know why you would want to turn this off… but you can.

- DV RX backlight turns the backlight on when a signal is received. Or not.

- TX Call Sign Display is an interesting one. You can elect to show My Call Sign, which is really your callsign while transmitting. Or Your Call Sign which is the call sign of the station that you are calling. I'm confused just writing it.

- Display Type (RX History) shows either Your name and callsign (which is the station that you are calling, or the repeater's name and callsign.

Sound

- Standby beep settings. Sets the beep when a received signal disappears. It can be set to OFF, ON, or two settings that only work with callsign routing.

- EMR (Enhanced monitor request) AF level. This is the equivalent of somebody shouting to get your attention.

Picture Settings ID-52

These settings are normally set on the radio or the Icom image management software, not in the CS program.

'Picture size' sets the size that an image will be sent. Images can be sent smaller than their native size, but not larger. You can also set the 'Picture quality' of the image. The default is 'Standard Quality 50%.'

The final setting is 'Receiver.' Any picture that you send can be received by any D-Star radio that has the picture-sharing capability. That could be anyone listening to your repeater, a linked gateway repeater, or a reflector. Entering a callsign here will indicate to any station that receives the image, who you intended it for, but it does not make it private.

GPS ID-52

This allows you to access the 300 GPS memories. They are mostly used for the alarm function. A beep is sounded when you receive a signal from within a GPS zone. There are many GPS-related options. All of this is covered very well in the Icom Advanced manual, so I won't repeat it here. Icom has dedicated 30 pages to GPS operation with the ID-52.

Common Settings ID-52

The common settings folder includes every menu setting available on the radio. Of course, some items will be changed as you operate the radio. This is a good reason to always load the current configuration from the radio into the CS program, before making any changes. If you rely on a version saved on your PC, you may unexpectedly change settings.

A/B Band Setting ID-52

The A/B Band Setting tab seems to contain everything that was not included elsewhere. Not just items relating to the A and B bands (VFOs).

It includes, Scan timers, scan skip settings, scan group links, Weather alert function (USA only), WX channel (USA only), and the Band Scope display mode. These settings would normally be changed on the radio, not in the CS software.

USING THE CS-51 SOFTWARE

You can use the CS-51 software to make changes to your radio configuration and to make a backup of the configuration if there is a problem. There are different versions of the CS-51 software for the ID-51, ID-51 plus and anniversary, and ID-51 plus 2.

TIP: Always accept the option of creating a backup on the SD card before performing a firmware update. Some updates delete all of your D-Star settings and saved memory entries.

Turn the radio on, connect the OPC-2350LU or OPC-2218LU data cable, and start the CS software. Or you can use the program to read a micro SD card that has been formatted in the radio. Once you have stored the updated configuration you can place the SD card back into the radio and restore the data.

The screen and features will be a little different for each ID-51 model. Annoyingly the right mouse click is not enabled on the CS-51 software.

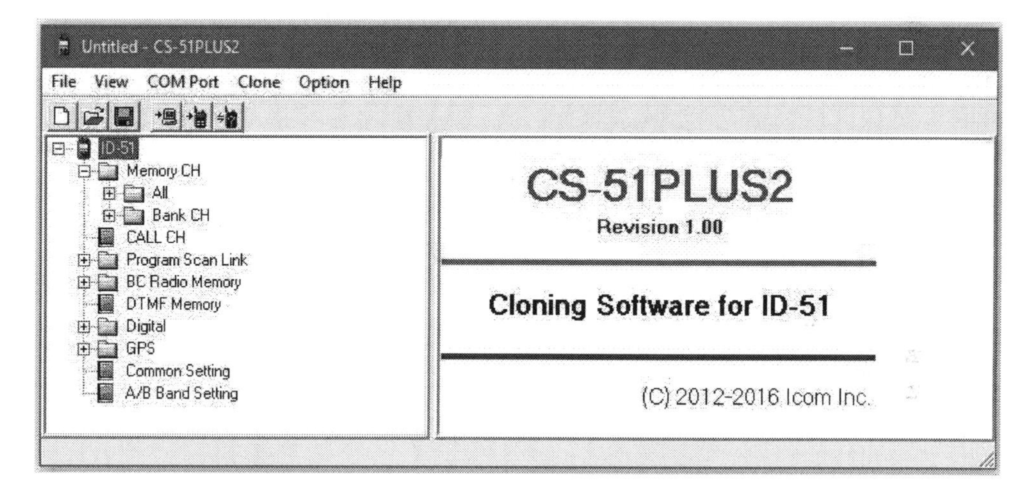

Figure 13: CS-51 plus 2 main screen

Memory CH folder ID-51

Opening the **Memory CH > All**, folder reveals 5 memory banks of 100 channels, making 500 channels in all. It is best to enter new channels into the Bank CH memory banks. Any channels you add there will automatically appear in the Memory Channel list, with the bonus of being arranged into groups.

You can set limits for up to 25 programmed scans in the **Memory CH > All > Program Scan Edge** table. For example, you might program a scan of the FM part of the 2m band, the whole 2m band, or the Aircraft band. The available modes are set according to the band of frequencies that you select. i.e., you can set AM for Aircraft band reception.

The **Scan Name** table lets you set names for the programmed scans.

The **Bank CH** folder holds 26 memory banks with 100 channels in each. Select the **Bank Name** item and add names for your memory banks. You always add more later. I have '2m FM,' '70cm FM,' 'Airband,' 'National' (repeater system), and 'DV' banks. The memory banks below 'Bank Name' will change to the names you have entered.

Select a group and enter appropriate memory channels into the list. Do not enter anything into the **CH Select** field. As soon as you enter a frequency, the program will pick the next available channel number. If you do enter a number into the field, the slot will populate with that channel from the main memory channel list. Even if it does not suit the current Bank name. Entering a new channel automatically adds the channel to the main memory channel list as well as the Bank CH list. The whole thing is really weird. Why not just have a bank number or letter on the main list and a short list of bank names?

TIP: If you enter a channel number a second time, for example in a different bank, the first one will disappear. The main purpose of the bank channels is for scanning. But the duplication seems unnecessary to me. This is implemented differently on the ID-52, where you can have the same channel in multiple banks.

You can click on a Bank name in the Bank CH list, or a channel group in the Memory Channel – All list and click **File > Import** or **File > Export** to import or export a .csv file into that particular list. The Memory CH All list has a field for Bank, so you can associate the channels into banks at the same time. I find the best thing s to export a list as a .csv file so that you know the correct headers. Add or edit channels using a spreadsheet, then import it back into the CS program, before finally uploading everything back to the radio.

CALL CH ID-51

There are four 'Call Channels,' two on the 2m band and two on the 70cm band. They are a quick way to get to channels that you use a lot, such as your local repeater or a simplex frequency that you use all the time.

Program Scan Link ID-52

This is almost too much flexibility. I doubt that very many people will ever get around to using this feature. I'm sure that I won't.

The idea is that you can create a scan list with gaps. The standard 'Program Scan Edges' scans all frequencies between upper and lower limits. The 'Program Scan Link' list, links two or more program scans together.

I guess it is useful, but I have all NZ FM repeaters in the memory channels, I could just scan that. There are ten **program scan links**, which can each be named in the **program scan link name table**.

BC Radio Memory ID-51

The BC Radio Memory can hold up to 500 broadcast channel frequencies in 26 groups. Each bank has a table for AM frequencies and one for FM frequencies. As with all the pages, you can import a list in .csv format.

DTMF Memory ID-51

The DTMF memory holds up to 16 DTMF short codes, used to control repeaters and link or unlink reflectors. You would probably normally program this on the radio rather than in the CS software. Each DTMF channel can hold 24 characters. See DTMF codes on page 46. Higher numbers in 'DTMF Speed' send the codes more slowly.

T-Call is only used on E-version radios, for European repeaters, it transmits a 1750 Hz 'in band' tone burst.

One channel can be marked with S. The DTMF code on that channel will be sent if you press the SQL key while transmitting.

Digital ID-51

The Digital folder holds the settings and lists for D-Star operation. Many of these settings and stores would normally be loaded with data while you are using the radio, rather than editing them in the CS software. This is another good reason to always load the current configuration from the radio into the CS program, before making any changes. If you rely on a version saved on your PC, you may unexpectedly change settings.

Since the callsign routing mode does not work, I use the rather misnamed '**Your Call Sign**' list to store reflectors that I want to access often in the DR mode. You can assign a name to each entry. I saved CQCQCQ, E, I, and U so that I can use the Dial to link to a reflector and then change to the CQCQCQ mode. The list can hold 200 entries.

The **repeater list** may already be populated on the radio. It can hold the worldwide list of D-Star repeaters. This enables the 'Near Repeater' function which lets you select repeaters that are near your current location according to fixed data or the GPS location. The list is supposed to be used for making routed 'Gateway CQ' calls to a remote repeater. But that does not usually work. Most D-Star users will never use the gateway mode. The ID-51 can store 750 repeaters.

My Station ID-51

The '**My Call Sign'** list can hold up to six callsigns for D-Star operation. It could be for six people that use your radio, or for six variants of your call sign. For example, ZL3DW / ID52, ZL3DW / SOTA, ZL3DW / MOB. The line marked with S for 'select' is the version that will be transmitted. You can change to a different version, on the radio at any time.

TX Message is a short message to be transmitted in the D-Star mode, along with your callsign. For example, "Portable at Lake Lindon." You can store up to six messages. The one with the S will be selected. You can turn the message feature on or off.

Transmitted call record ID-51

The transmitted call record is called TX History on the newer radios. It stores the repeaters you have used most recently, who you called on FM (in DR mode), who you called on DV (in DR mode), and the reflectors you connected to. This screen is not intended to be edited with the CS software, but you could use it as a short-term memory bank.

Received call record ID-52

The received call record is called RX History on the newer radios. It shows the history of received calls. It is not intended to be edited with the CS software. 'Rx RPT1' is the gateway the call came through, usually, your local repeater and 'Rx RPT2' is the D-Star identity of the repeater. If you are using a hotspot, 'Rx RPT1' will probably be your call with a G module and 'Rx RPT2' will probably be your call with a band module (A, B, C).

Digital Setting ID-51

This page sets all the D-Star digital settings on the radio.

Tone Controls

- Tone controls for received audio

DV Set Mode

- Auto Reply sets an automatic voice message reply to a call made to your callsign. The 'Position' setting is a great safety feature. If someone calls you, the radio responds with your call sign and location. So, if you were injured on a trail or something like that, the radio would send your location. It could also be useful if you were mobile and could not talk on the radio immediately. Sadly, Auto Reply is unlikely to work because it relies on callsign routing.

- The DV Data TX function sets whether data is sent immediately after it is received from a connected device such as a PC or a weather station etc. Or if it has to wait until you press the PTT.

- DV Fast Data TX Delay (PTT), sets what happens if you press the PTT while the radio is sending fast data (usually a picture). The function only works if The DV Data TX function is set to PTT.

  If DV Fast Data TX Delay (PTT) is set to OFF, the radio switches to slow data while the PTT is pressed and then back to receive when you release the PTT.

The remaining data will continue to be sent in the slow picture mode each time you transmit.

If the item is set between **1 and 10 seconds**, the radio switches to slow data while the PTT is pressed and then back to fast data for the nominated time when you release the PTT. If all the data has been sent, the radio returns to receiving. After the nominated period, the radio returns to receiving anyway.

- Digital Monitor can be set to listen to the channel in DV or FM, or automatically switch between DV and FM.
- Digital Repeater Set, selects whether you want the repeater callsign to be corrected in the event that it is different to the setting you have in the FROM box. I can't see any point in turning this off.
- RX callsign auto write. If someone sends a call to your callsign the radio automatically puts their callsign into the UR field (TO box). Seems like a good idea to me.
- Repeater callsign auto write. If someone sends a call to your callsign and your radio is not in the DR mode, the radio automatically sets the R1 and R2 repeater callsigns.
- DV mode auto detect, switches the radio to DV if a digital signal is received and to FM if an FM signal is received. Yep, that sounds pretty useful! Leave it switched on.
- RX Record (RPT). The transceiver can record up to 50 calls. You can select whether to record all calls or only the most recent (current) one.
- **RX->CS** key. If this is set to **Call Sign Capture**, holding down the **RX->CS** key will store the current (last) callsign into the RX History. If you rotate the **Dial** knob while holding down **RX->CS**, you can select a station to call. If this is set to **RX->CS list**, rotate the **Dial** while holding down **RX->CS**, to select a station to call. This requires callsign routing, so it probably won't work.

Display

- RX Call Sign Display selects what happens when you receive a signal. In the 'Auto' mode the calling station's callsign and message are displayed once. In the 'Auto RX Hold' mode, the callsign and message are displayed and then the callsign remains displayed. I find this very useful as I always forget the calling station's callsign. The 'Hold' mode is the same as the 'RX Hold' mode except the message and callsign continue to scroll every two seconds until another call is received.
- Reply Position Display sets whether a signal sent back via the auto-reply mode will show the other station's position. Since this is pretty much the main

reason to use the auto-reply feature, I would expect that you would leave this turned on. This requires callsign routing, so it probably won't work.

- DV RX backlight turns the backlight on when a signal is received. Or not.
- TX Call Sign Display is an interesting one. You can elect to show **My Call Sign** which is really your callsign while transmitting. Or **Your Call Sign** which is the call sign of the station that you are calling. I'm confused just writing it.
- Display type (DR). Select a large or normal name display
- Display Type (RX History) shows either **Your** (UR) name and callsign which is the station that you are calling, or the repeater's name and callsign.

Sound

- Standby beep settings. Sets the beep when a received signal disappears. It can be set to OFF, ON, or two settings that only work with callsign routing.
- EMR (Enhanced monitor request) AF level. This is the equivalent of somebody shouting to get your attention.

GPS ID-51

This allows you to access the 200 GPS memories in 26 banks. They are mostly used for the alarm function. A beep is sounded when you receive a signal from within a GPS zone. There are also many GPS-related options which are covered very well in the Icom Advanced manual, so I won't repeat them here.

Common Settings ID-51

The common settings folder includes every menu setting available on the radio. Of course, some items will be changed as you operate the radio. This is a good reason to always load the current configuration from the radio into the CS program, before making any changes. If you rely on a version saved on your PC, you may unexpectedly change settings.

A/B Band Setting ID-51

The A/B Band Setting tab includes Scan timers, scan skip settings, and scan group bank link activating switches.

USING THE CS-705 SOFTWARE

The CS-705 software is for the IC-705. You can use the CS software to make changes to your radio configuration and to make a backup of the configuration if there is a problem.

TIP: Always accept the option of creating a backup on the SD card before performing a firmware update. Some updates delete all of your D-Star settings and saved memory entries.

Turn the radio on, connect the USB cable as described above, and start the CS software.

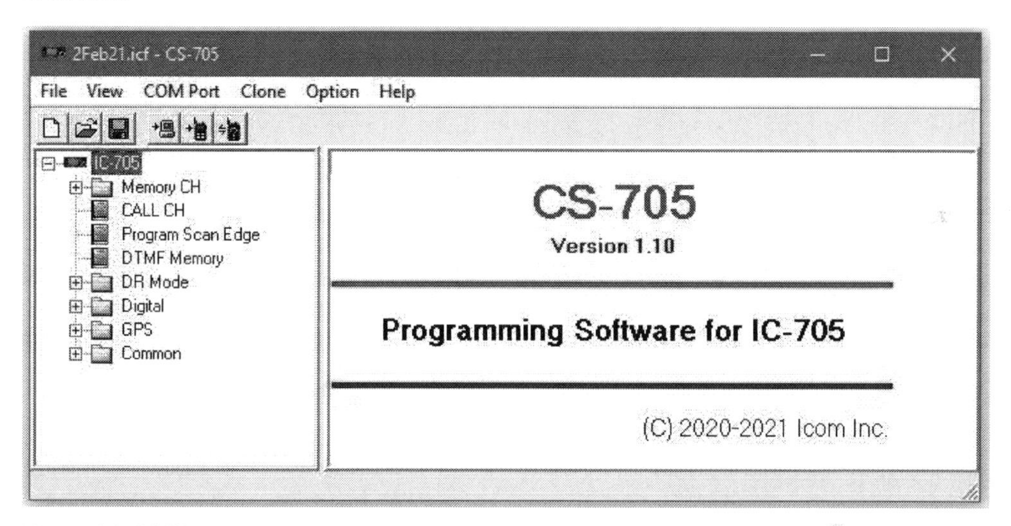

Figure 14: CS-705 main screen

Memory CH folder IC-705

Opening the memory channel folder reveals up to 100 groups. Each group can hold up to 100 stored memory channels, with a total memory capacity of 500 'normal' channels.

Channels can be in more than one group. For example, you might have a group with all New Zealand channels, another for Local channels, and another for 70cm repeaters. Some channels might be in all three groups.

Right-click a folder to import a group from a .csv formatted data file. It is best to enter one channel and use **right-click> export** to export a .csv file to your computer. That will provide the column headers for the repeater file that you wish to import. You can import a .csv file which includes multiple groups.

CALL CH IC-705

There are four 'Call Channels,' two for the 2m band and two for the 70cm band. They are a quick way to get to channels that you use a lot, such as your local repeater or a simplex frequency that you use all the time.

Program Scan Edge IC-705

You can set limits for up to 25 programmed scan ranges. They can be on the HF bands or the VHF and UHF bands. For example, you might program a scan of the FM part of the 2m band, the whole 2m band, the Aircraft band, or the FM Marine band.

The IC-705 is an all-mode radio so any mode can be selected for a particular scan range. Data mode can be activated for modes that support data transmission FM-D, USB-D etc. You can also select the receiver filter bandwidth 1, 2 or 3, to match the selected mode.

DTMF Memory IC-705

The DTMF memory holds up to 16 DTMF short codes, used to control repeaters and link or unlink reflectors. You would probably normally program this on the radio rather than the CS software. Each DTMF channel can hold 24 characters. See DTMF codes on page 46. You can set the speed of the DTMF transmission. Higher numbers in 'DTMF Speed' send the codes more slowly.

DR Mode IC-705

The DR Mode folder holds information relating to D-Star operation. Nothing on this screen is very important.

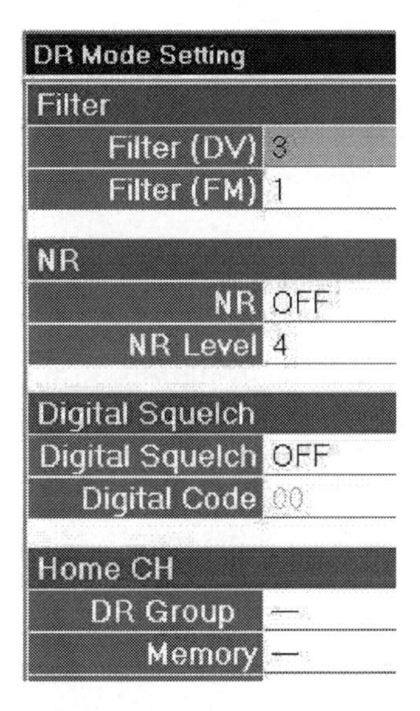

Figure 15: Typical IC-705 DR mode setting options

The **filter** selection applies to all stored DR repeaters. I set the DV filters to **3** which is an IF bandwidth of 7 kHz. That should be adequate for the 6 kHz wide D-Star signal.

I don't usually store FM channels in the DR memory, but I did put my two local FM repeaters in because it means you don't have to leave the DR mode to quickly check them for activity. I set the FM filters to **1** which is an IF bandwidth of 15 kHz.

I don't set the **NR** (noise reduction) settings. They are normally not required for FM channels and are never used for DV channels. **Digital squelch** is not commonly used. It only applies to the FM channels. Digital Voice signals are squelched according to whether data is being received or not.

The **Home CH** setting is not very important. You can change a menu setting so that a beep sounds when you select the channel in the FROM box in the DR mode. "Yay!" You can also program one of the preset buttons or one of the microphone buttons to provide a quick change to the nominated 'home' channel. It's not something that I have bothered to set up. This setting does not affect the channel that becomes active when you select the DR mode. Other than immediately after a reset, the radio always returns to the last used FROM and TO settings.

Digital IC-705

Since the callsign routing mode does not work, I use the rather misnamed '**Your Call Sign**' list to store reflectors that I want to access while in the DR mode. You can assign a name to each entry.

Your Call Sign (Remain 189 memories)

No.	Name	Call Sign
1	Unlink	U
2	Echo	E
3	Info	I
4	Use Reflector	CQCQCQ
5	REF001 C	REF001CL
6	REF030 C	REF030CL
7	Kiwi chat	XLX299KL
8	WW 299	XLX299AL
9	BM530	XLX299BL
10	QuadNet Tech	XRF757CL
11	QuadNet Array	XRF757AL
New		

Figure 16: The IC-705 'Your Call Sign' list can hold the reflectors you like to use.

I saved CQCQCQ, E, I, and U as well so that I can use the **Dial** to link to a reflector and then change to the CQCQCQ mode. The list can hold 300 entries. Note that E, I, and U have seven preceding spaces.

The **repeater list** may already be populated on the radio. It is designed to hold the worldwide list of D-Star repeaters, but you only need repeaters that you intend to use and your hotspot, if you have one. The latitude and longitude information enables the 'Near Repeater' function which lets you select repeaters that are near your current location according to fixed data or the GPS location. The list is supposed to be used for making routed 'Gateway CQ' calls to a remote repeater. But that does not usually work. Most D-Star users will never use the gateway mode. The IC-705 can store 2500 repeaters.

My Station IC-705

The '**My Call Sign'** list can hold up to six callsigns for D-Star operation. It could be for six people that use your radio, or for six variants of your call sign. For example, ZL3DW / 705, ZL3DW / SOTA, ZL3DW / MOB. The line marked with S for 'select' is the version that will be transmitted. You can change to a different version, on the radio at any time.

TX Message is a short message to be transmitted in the D-Star mode, along with your callsign. For example, "Andrew Christchurch." You can store up to six messages. The one with the S will be selected. You can turn the message feature on or off.

TX History IC-705

The TX History stores the repeaters you have used most recently, who you called, and the reflectors you connected to. This screen is not intended to be edited with the CS software, but you could use it as a short-term memory bank.

RX History IC-705

The RX History shows the history of received calls. It is not intended to be edited with the CS software. 'Rx RPT1' is the gateway the call came through, usually, your local repeater and 'Rx RPT2' is the D-Star identity of the repeater. If you are using a hotspot, 'Rx RPT1' will probably be your call with a G module and 'Rx RPT2' will probably be your call with a band module (A, B, C).

Digital Settings IC-705

This page sets all the D-Star digital settings on the radio.

DV Set Mode

- Automatic reply to a call made to your callsign. Most folks hang out on the reflectors, but this can be handy if you have a buddy that calls you direct. The 'Position' setting is a great safety feature. If someone calls you, the radio responds with your call sign and location. So, if you were injured on a trail or something like that, the radio would send your location. It could also be useful if you were mobile and could not talk on the radio immediately. This function relies on callsign routing which probably won't work.

- The DV Data TX function sets whether data is sent immediately after it is received from a connected device such as a PC or a weather station etc. Or if it has to wait until you press the PTT.
- DV Fast Data TX Delay (PTT), sets what happens if you press the PTT while the radio is sending fast data (usually a picture). The function only works if The DV Data TX function is set to **PTT**.

  If DV Fast Data TX Delay (PTT) is set to **OFF**, the radio switches to slow data while the PTT is pressed and then back to receive when you release the PTT. The remaining data will continue to be sent in the slow picture mode each time you transmit until the image has all been sent.

  If the item is set between **1 and 10 seconds**, the radio switches to slow data while the PTT is pressed and then back to fast data for the nominated time when you release the PTT. If all the data has been sent, the radio returns to receiving. After the nominated period, the radio returns to receiving anyway.
- Digital Monitor can be set to listen to the channel in DV or FM, or automatically switch between DV and FM.
- Digital Repeater Set, selects whether you want the repeater callsign to be corrected in the event that it is different to the setting you have in the FROM box. I can't see any point in turning this off.
- DV mode auto detect, switches the radio to DV if a digital signal is received and to FM if an FM signal is received. Yep, that sounds pretty useful! Leave it switched on.
- RX Record (RPT). The transceiver can record up to 50 calls. You can select whether to record all calls or only the most recent (current) one.

Display

- RX Call Sign Display selects what happens when you receive a signal. In the 'Normal' mode, the calling station's callsign and message are displayed once. In the 'RX Hold' mode, the callsign and message are displayed and then the callsign remains displayed. I find this very useful as I always forget the calling station's callsign. The 'Hold' mode is the same as the 'RX Hold' mode except the message and callsign continue to scroll every two seconds until another call is received.

- The 'RX Position Indicator' setting controls whether the radio will display received position information. The 'RX Position Display Timer' sets how long the position will be displayed.
- Reply Position Display sets whether a signal sent back via the auto-reply mode will show the other station's position. Since this is pretty much the main reason to use the auto-reply feature, I would expect that you would leave this turned on. This mode requires callsign routing, so it probably won't work.
- RX picture indicator sets whether the received picture icon will be displayed when a picture is received. I don't know why you would want to turn this off... but you can.
- DV RX backlight turns the backlight on when a signal is received. Or not.
- TX Call Sign Display is an interesting one. You can elect to show **My Call Sign** which is really your callsign, while transmitting. Or **Your Call Sign** which is the call sign of the station that you are calling. I'm confused just writing it.
- Display Type (RX History) shows either **Your** (UR) name and callsign which is the station that you are calling, or the repeater's name and callsign.

Sound

- Standby beep settings. Sets the beep when a received signal disappears. It can be set to OFF, ON, or two settings that only work with callsign routing.
- EMR (Enhanced monitor request) AF level. This is the equivalent of somebody shouting to get your attention.

DV Gateway IC-705

This page allows you to choose between a WiFi connection or a USB connection for establishing Terminal Mode connections. Select **Global** if you are outside Japan. You can set a callsign that is allowed access via the Access Point mode.

For more information, you can download a 29-page manual from Icom called, 'About the DV Gateway function.' You will have to open port forwarding of port 40000 on your home router to enable communication to the linked gateway. You also need to know the IP address of a G3 D-Star gateway that is using the RS-RP3C program.

I believe that it is much easier to abandon the Terminal Modes and use an MMDVM hotspot instead. Hotspots are easier to use, and you do not have to open a port on your router to the outside world.

Picture Settings IC-705

These settings are normally set on the radio or the Icom image management software, not in the CS program.

'Picture size' sets the size that an image will be sent. Images can be sent smaller than their native size, but not larger. You can also set the 'Picture quality' of the image. The default is 'Standard Quality 50%.'

The final setting is 'Receiver.' Any picture that you send can be received by any D-Star radio that has the picture-sharing capability. That could be anyone listening to your repeater, a linked gateway repeater, or a reflector. Entering a callsign here will indicate to any station that receives the image, who you intended it for, but it does not make it private.

GPS IC-705

This allows you to access the 300 GPS memories. They are mostly used for the alarm function. A beep is sounded when you receive a signal from within a GPS zone. There are also many GPS-related options. All of this is covered very well in the Icom Advanced manual, so I won't repeat it here. Icom has dedicated 46 pages to GPS operation with the IC-705.

Common Settings IC-705

The common settings folder contains many screens which include every menu setting available on the radio. Of course, some items will be changed as you operate the radio. This is a good reason to always load the current configuration from the radio into the CS program, before making any changes. If you rely on a version saved on your PC, you may unexpectedly change settings.

The **WiFi** can be set to connect to a WiFi network or to become an access node that you can connect your phone to. This can be used to send photos from your phone to the radio for D-Star transmission when you are operating portable from some lonely mountaintop.

There is a sub-folder for settings related to the new **Preset** function. The radio can store up to five different presets.

Automatic repeater edges. The automatic repeater settings should be preset for your region when you get the radio, but there may be some differences in your local band plan. You can assign frequency ranges where the radio will assume that you are tuned to repeater outputs. Within those ranges, you can set whether the radio will apply a positive offset or a negative offset, (DUP+ or DUP-). For example, on the 2m band in New Zealand, repeaters with output frequencies between 146.100 MHz and 146.999 MHz require a negative 600 kHz split. Your radio transmits lower than the receiver frequency. Repeaters with output frequencies between 147.000 MHz and 147.499

MHz require a positive 600 kHz split. Your radio transmits higher than the receiver frequency.

USING THE CS-9700 SOFTWARE

The CS-9700 software for the IC-9700 is similar to the IC-705 version. It has a few more features due to the multi-band and satellite configuration options. You can use the CS software to make changes to your radio configuration and to make a backup of the configuration if there is a problem.

TIP: Always accept the option of creating a backup on the SD card before performing a firmware update. Some updates delete all of your D-Star settings and saved memory entries.

Turn the radio on, connect the USB cable as described above, and start the CS software.

I recommend that you take a copy of the existing setup, before making any changes. It ensures that you are changing what is actually on the radio, not some previous iteration, and it serves as a backup in case of trouble. It also confirms that the software can communicate with the radio.

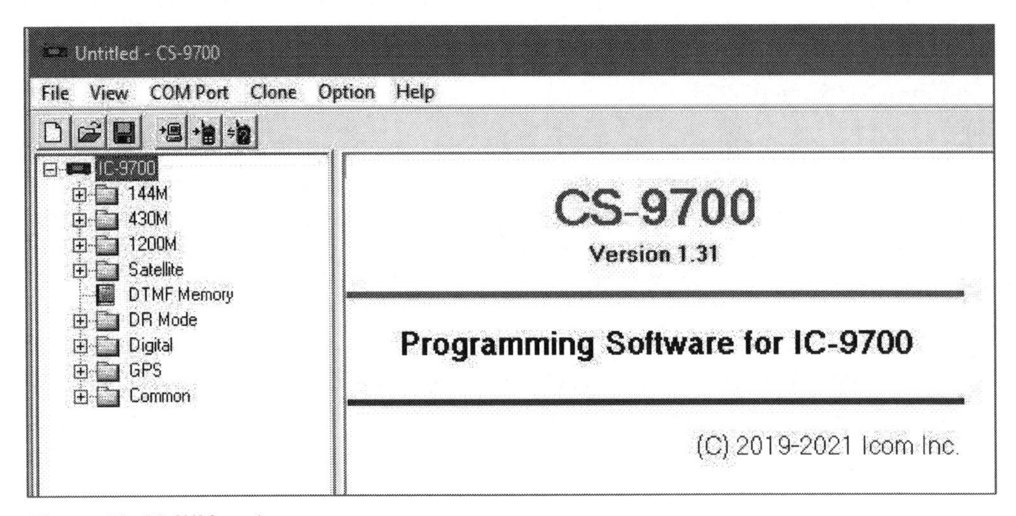

Figure 17: CS-9700 main screen

Reading and writing data

Click the '**Clone Read**' icon. It is the one that shows an arrow pointing at a computer screen. After the data has been transferred, hit the **Save** icon and save the file to your PC. Now you can make changes. After that, **Save** the file again with a different file name and click on the '**Clone Write**' icon. It is the one with an arrow pointing at a handheld radio.

Do not disconnect the radio while data is being read from or written to the radio.

FM folders IC-9700

The 144M folder holds up to 99 stored memory channels for the 2m band. You can store FM and DV channels, but my preference is to only use this memory bank for FM channels. I use the DR memory for D-Star channels. You can add, edit, or delete channels using the CS program, or on the radio.

The 430M folder holds up to 99 stored memory channels for the 70cm band. You can store FM and DV channels, but my preference is to only use this memory bank for FM channels.

The 1200M folder holds up to 99 stored memory channels for the 23cm band. You can store FM, DV, and DD (digital data) channels.

The Satellite folder holds satellite uplink and downlink channels. Usually, 2m and 70cm crossband.

Notes: double click SEL to set the IF filter bandwidth. DV and FM have a choice of three non-adjustable filters. The DD mode is fixed at 130 kHz. The other modes have a choice of three adjustable filter settings. +DUP means that the radio will transmit higher by the Offset Frequency. -DUP means that the radio will transmit lower by the Offset Frequency. Check with your repeater band plan. TONE means that the radio will transmit a tone to open the FM repeater squelch. The TSQL mode squelches your receiver until the correct tone is received from the repeater. It also sends the tone on your transmission to open the repeater squelch. The Repeater Tone is usually the same as the TSQL tone. If the tone is set to OFF or omitted, the squelch will open when the received level is stronger than the squelch level. No tone is required for receiving and no tone is transmitted. Digital squelch is not commonly used.

For each band you can set, program scan edges, calling channels, edit the band stacking register, edit the memo pad, change the filters for each mode, set power limits, and adjust noise filters, RIT etc. (although these will be overwritten when you make changes on the radio). The satellite folder has its own band stacking register.

DTMF IC-9700

The DTMF memory holds DTMF short codes, used to control repeaters and link or unlink reflectors. You would probably normally program this on the radio rather than using the CS software. See DTMF codes on page 46. Higher numbers in 'DTMF Speed' send the codes more slowly. I had to change to 200ms.

The usable DTMF codes are 0 to 9, A to D, *, and #.

I have d0 set to 0, which resets a repeater to its default link. It might be 'unlinked.'

d1 is set to #, which is the same as the U command (Unlink reflector).

d3 is set to **, which is the same as the I command (Information).

DR mode folder IC-9700

The DR Mode folder holds information relating to D-Star operation. Nothing on this screen is very important.

The **filter** selection applies to all stored DR repeaters. I set the DV filters to **3** which is an IF bandwidth of 7 kHz. That should be adequate for the 6 kHz wide D-Star signal.

I don't usually store FM channels in the DR memory, but I did put my two local FM repeaters in because it means you don't have to leave the DR mode to quickly check them for activity. I set the FM filters to **1** which is an IF bandwidth of 15 kHz.

I don't set the **NR** (noise reduction) settings. They are normally not required for FM channels and are never used for DV channels. The same applies to **RIT** settings, which only apply to SSB operation. **Digital squelch** is not commonly used. It only applies to the FM channels. Digital Voice is squelched according to whether data is being received or not.

The Home CH setting is not very important. You can change a menu setting so that a beep sounds when you select the channel in the FROM box in the DR mode. "Yay!" You can also program one of the preset buttons or one of the microphone buttons to provide a quick change to the nominated 'home' channel. It's not something that I have bothered to set up. This setting does not affect the channel that becomes active when you select the DR mode. Other than immediately after a reset, the radio always returns to the last used TO and FROM settings.

DR Mode Setting

Filter	DV	FM
144M	3	1
430M	3	1
1200M	3	1
NR	**MAIN**	**SUB**
NR	OFF	OFF
NR Level	4	4
RIT	**MAIN**	**SUB**
RIT	OFF	OFF
RIT Frequency	0.000kHz	0.000kHz
Digital Squelch	**MAIN**	**SUB**
Digital Squelch	OFF	OFF
Digital Code	00	00
Home CH		
DR Group	01: NZ	
Memory	Pi Star	
	ZL3DW B	438.125000 DV

Figure 18: Typical IC-9700 DR mode setting options

Digital IC-9700

Since the callsign routing mode does not work, I use the rather misnamed '**Your Call Sign**' list to store the reflectors that I want to access often in the DR mode. You can assign a name to each entry. I saved CQCQCQ, E, I, and U as well so that I can use the **Dial** to link to a reflector and then change to the CQCQCQ mode. The list can hold 300 entries. Note that E, I, and U have seven preceding spaces.

The **repeater list** may already be populated on the radio. It is designed to hold the worldwide list of D-Star repeaters, but you only need repeaters that you intend to use and your hotspot, if you have one. The latitude and longitude information enables the 'Near Repeater' function which lets you select repeaters that are near your current location according to fixed data or the GPS location. The list is supposed to be used for making routed 'Gateway CQ' calls to a remote repeater. But that does not usually work. Most D-Star users will never use the gateway mode. The IC-9700 can store 2500 repeaters.

Your Call Sign (Remain 189 memories)

No.	Name	Call Sign
1	Unlink	U
2	Echo	E
3	Info	I
4	Use Reflector	CQCQCQ
5	REF001 C	REF001CL
6	REF030 C	REF030CL
7	Kiwi chat	XLX299KL
8	WW 299	XLX299AL
9	BM530	XLX299BL
10	QuadNet Tech	XRF757CL
11	QuadNet Array	XRF757AL
New		

Figure 19: The IC-9700 'Your Call Sign' list can hold the reflectors you like to use.

GPS FOLDER IC-9700

The GPS folder holds the 300 GPS memories which can be saved into 26 groups. Each group can have a name. The GPS memories are primarily used for the GPS Alarm Function. I don't use this function, so on my radio, it contains default data such as the locations of the three largest Ham Fests, Tokyo, Dayton, and Friedrichshafen. Plus, Icom HQ and some other locations in Japan.

You can save your current position, another station's position, or a manually entered position using the menu commands on the radio. The GPS folder also includes the **GPS setup page** which has the port settings for the external GPS receiver. If you change **GPS Select** to **Manual**, you can manually enter your location.

Compass direction sets 'North up,' 'South up,' or your current 'Heading up.' You can change whether to display received GPS data from the main receiver, the sub-receiver, or the latest received signal irrespective of the band. You can change from transmitting D-PRS data (default), or NMEA data, and set a beacon timer that transmits your location at regular intervals.

The remaining settings control what information is sent. You can change the D-PRS icon from, a house, to a car, or a person walking etc. You can connect to a weather station and send weather station data. If you elect to send NMEA data, you can change which NMEA sentences are transmitted.

All of this is covered very well in the Icom Advanced manual, so I won't repeat it here. Icom has dedicated 43 pages to GPS operation with the IC-9700.

Common Folder IC-9700

The common folder contains several screens which include every setting available on the radio. Of course, some items will be changed as you operate the radio. You might change a filter setting or something. This is another good reason to always load the current configuration from the radio into the CS program, before making any changes. If you rely on a version saved on your PC, you may unexpectedly change settings.

Memory Keyer. This is the best place to enter the eight memory keyer strings for the Morse Code keyer. The **Keyer** tab includes the keying speed, contest number, sidetone level, CW waveform adjustments, and Break-in settings. Some settings will change as you use the radio.

The radio includes an RTTY memory keyer. You can enter the eight memory keyer strings on the **RTTY TX Memory** tab.

There are settings for the **Connectors** and the **Network** (Ethernet port). The **Scope** and **Audio Scope** tabs enable you to change the band scope colours.

And there is a sub-folder for settings related to the new **Preset** function. The radio can store up to five different presets.

Automatic repeater edges. The automatic repeater settings should be preset for your region when you get the radio, but there may be some differences in your local band plan. You can assign frequency ranges where the radio will assume that you are tuned to repeater outputs. Within those ranges, you can set whether the radio will apply a positive offset or a negative offset, (DUP+ or DUP-). For example, the New Zealand 2m band repeaters with output frequencies between 146.100 MHz and 146.999 MHz require a negative 600 kHz split. Your radio transmits lower than the receiver frequency. Repeaters with output frequencies between 147.000 MHz and 147.499 MHz require a positive 600 kHz split. Your radio transmits higher than the receiver frequency.

GPS location

GPS location information is used for three things. It can be transmitted with your D-Star transmissions so that the station you are calling can see; your location, Maidenhead grid reference, distance, speed, the direction you are heading, and what direction you are from them. It is also used for the 'Near Repeater' function which lets you choose DV or FM repeaters that are near your location. And it is used to determine the distance and heading towards stations that you hear via the repeater or gateway.

If you want to run D-Star but you don't plan to operate mobile or portable, there is little point in connecting an external GPS to a radio that doesn't have a GPS receiver built-in. Just enter your location manually. The IC-705 and ID-52 have a built-in GPS receiver, but if you are operating indoors, you might still want to load a manual position.

Connecting a GPS receiver to an IC-9700

If you plan on operating in D-Star mode while mobile or portable, you can connect the radio to an external GPS unit. It has to be NMEA compatible, and it has to output at RS-232 data levels. The expected data rate is selectable, but the default is 9600 bps. The GPS receiver is connected to the 2.5mm 'Remote' jack on the rear panel.

The TTL data level on most of the u-blox NEO receivers available online is too low. But you can buy a TTL to RS232 adaptor board for about US $2.

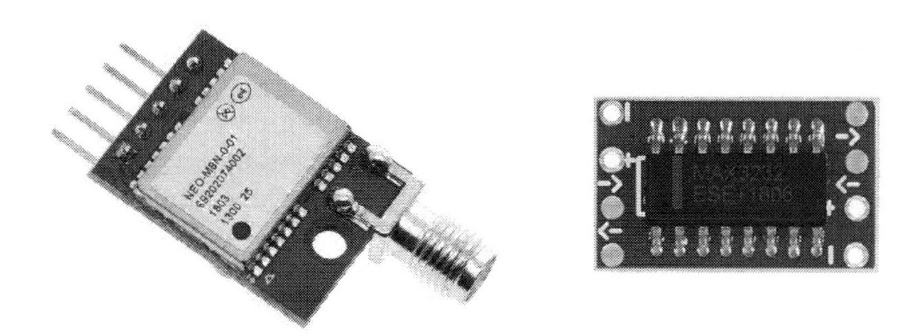

Figure 20: Typical GPS receiver board and RS232-TTL adapter

Or you can buy a GPS board that already uses RS232 levels. I ended up buying a Beitian BS-280 module from the Alexnld.com Internet site. It outputs NMEA-0183 data at 9600 bauds at RS-232 levels and it cost 18 USD including postage to New Zealand. The unit requires a 5V DC supply. I powered mine from a USB hub.

The connection to the radio is a 2.5 mm stereo mini phono connector. RX-D (green) on the GPS is connected to the ring, TX-D (white) is connected to the tip and the earth is the black wire.

Set **Menu > GPS > GPS Set**, to **External GPS** and the baud rate to suit the GPS unit. It is 9600 baud for the Beitian and usually 4800 for u-blox NEO boards.

When you return to the main radio screen, you should see a little satellite icon at the top of the screen. It will flash if it is receiving data but has not yet seen enough satellite data. It stops flashing after data has been received from several satellites. This may take several minutes. If you don't see the satellite icon, the chances are that you have the RX-D and TX-D connection reversed, or the level from the unit is not sufficient.

TIP: The GPS receiver will also keep the clock display accurate. Select ***MENU > SET > Time Set > Date/Time > GPS Time Correct > Auto.***

GPS Information IC-705 and IC-9700

QUICK > GPS Information is a quick way of displaying the GPS information screen. It is the same screen as you get through the **Main > GPS** sub-menu, but it is easier to access from the QUICK button. It will only be displayed on the IC-9700 if you have an external GPS receiver connected to the data input.

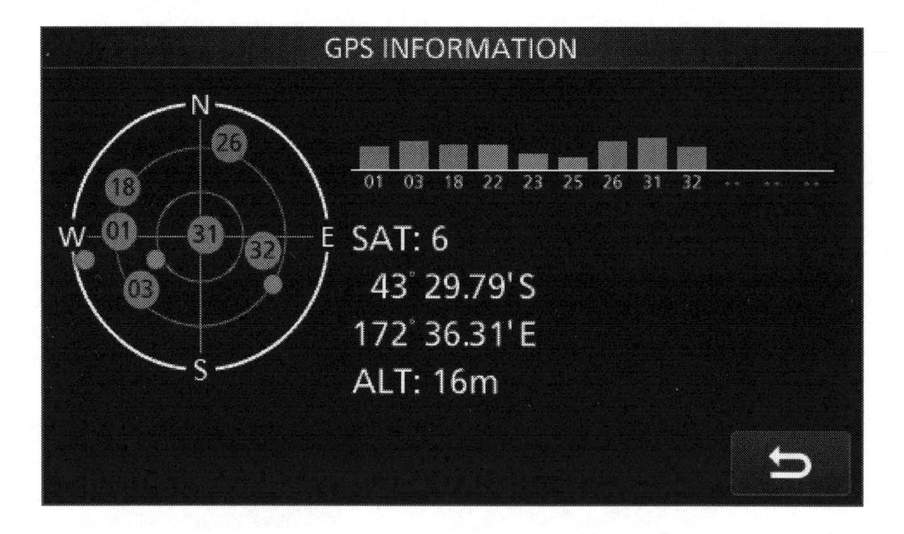

The GPS Information screen shows the satellite number and bearing of satellites being received by the connected GPS receiver. The image is a 'Radar Plot.' The centre is your location. Satellites that are high overhead are shown close to the centre. The outer ring is the horizon. The white dots are GPS satellites that are being received but the data is invalid, usually because of extreme range or some path obstruction.

GPS Position IC-705 and IC-9700

QUICK > GPS Position is a quick way of displaying the four GPS position screens. It is the same screen as you get through the **Main > GPS** sub-menu but it is easier to access from the QUICK button.

MY: The first GPS Position screen shows your location. Latitude, longitude, Maidenhead Grid, altitude (can be inaccurate), your speed, course, and the current time.

TIP: If the Maidenhead Grid locator is radically different to what you expected, you probably have the manual setting of the East/West or North/South setting wrong.

RX: The second screen shows the GPS data from a received D-Star signal.

MEM: The third screen shows the GPS data from a location stored in the GPS memory bank. Touch and hold the screen to load a location from the GPS memory.

ALM: The fourth screen can show the GPS Alarm distance and bearing to a nominated location. When the alarm is set, a beep is issued by the radio when you drive within 1 km of the nominated site or group of GPS locations. At 500 metres three beeps are issued. This is a function known as 'geo-fencing,' (advanced manual pages 9-26). It can also be used to alert you when a 'target' station in your list comes within 1 km or 500 metres of your location or the selected area. The station has to transmit within the area for the alarm to work.

GPS Information ID-52 and ID-51 +2

`QUICK > GPS Information` shows a radar "Sky view" image of the received GNSS (global navigation satellite system), satellites. The centre is your location. Satellites that are high overhead are shown close to the centre. The centre is 90 degrees in elevation. The middle rings indicate 60 degrees of elevation and 30 degrees of elevation. The outer ring is the horizon at 0 degrees of elevation. The grey dots are GPS satellites that are being received but the data is invalid, usually because of extreme range or an obstructed path.

Light blue blobs indicate weak satellite signals. Dark blue blobs indicate strong signals. The number in the blob is the satellite's 'PRN' number.

GPS satellites have numbers between 1 and 32. SBAS satellites are mostly geostationary satellites that improve the accuracy of the GPS service. Their numbers range from 33 to 71. GLONASS (Russian GNSS) satellites have numbers between 65 and 96 and QZSS (Japanese GNSS) satellites have numbers between 193 and 202.

The information screen also indicates your altitude, latitude, longitude, and the number of satellites being received.

GPS Position ID-52 and ID-51

The ID-52 and ID-51 can display four sets of GPS position information. The data provided is dependent on what was sent from the other station. The position information that you transmit is also customisable.

Press QUICK > GPS Position > rotate the Dial to select from MY, RX, MEM, or ALM.

TIP: IF you transmit while this screen is open it will close. But you can open it again while you are transmitting.

MY: The first GPS Position screen shows your location. latitude, longitude, Maidenhead Grid, altitude (can be inaccurate), your speed, compass bearing, and the current time.

TIP: If the Maidenhead Grid locator is radically different to what you expected, you probably have the manual setting of the East/West or North/South setting wrong.

RX: The second screen shows the GPS data from a received D-Star signal. The direction towards the station is shown on a compass. You also get their grid square, which is useful for your log and some awards. And their latitude and longitude, altitude, direction in degrees East of North, and speed. And finally, the time that the signal was received. The ID-51 splits this information over two screens.

MEM: The third screen shows the GPS data from a location stored in the GPS memory bank. Touch and hold the screen to load a location from the GPS memory.

ALM: The fourth screen can show the GPS Alarm distance and bearing to a nominated location. When the alarm is set, a beep is issued by the radio when you drive or walk within 1 km of the nominated site or group of GPS locations. At 500 metres three beeps are issued. This is a function known as 'geo-fencing,' (see 'GPS Alarm' in section 6 in the Icom advanced manual). It can also be used to alert you when a 'target' station in your list comes within 1 km or 500 metres of your location or a nominated area. The station has to transmit within the area for the alarm to work.

ENTERING YOUR LOCATION MANUALLY

The first thing you need to know is the latitude and longitude of your location. You can get this from a phone App like 'Maidenhead,' Google Earth or Google Maps (click on the map and look at the URL line).

Manual position on the IC-705 and IC-9700 and similar radios

To enter your location manually select MENU > 2 > GPS > GPS Set > Manual Position. Touch and hold anywhere on the screen, then select Edit.

TIP: If you have a GPS receiver connected and External GPS selected, you can load the manual data from that, or you can load it from a saved GPS memory.

Touch 'Latitude' and enter the latitude. If you live in the southern hemisphere, remember to change N to S using the N/S Soft Key. Touch ENT to save and exit.

Touch 'Longitude' and enter the longitude. Remember to set the East-West setting using the E/W Soft Key. Touch ENT to save and exit.

TIP: Since firmware update 1.20 you can also enter latitude and longitude as degrees with decimal points.

Touch Altitude and set your altitude above sea level if you know it. Or just set it to zero. Touch ENT to save and exit.

Manual position on the ID-52 and similar radios

Normally you would just use the position provided by the built-in GPS receiver. But if you are inside where there is no GPS reception you can set your position manually. To enter your location manually select MENU > GPS > GPS Set > GPS Select > Manual Position. Then press QUICK and select Edit or capture from GPS.

Set MENU > GPS > GPS Set > GPS Select to Manual to use the manual setting rather than the received GPS location.

GPS LOGGER FUNCTION

The GPS Logger function records your location at preset intervals so that you can plot the route that you took by importing the data into mapping software. The feature is available on the ID-31, ID-51, ID-52 and IC-705 because they are portable radios. The GPS Logger saves Latitude, Longitude, Altitude, Positioning state, Course, Speed, Date, and Time onto the SD card. Icom has issued a useful document called 'Using the GPS Logger.' The altitude function is great for SOTA (summits on the air) operation. Set the interval using Menu > GPS > GPS Logger > Record Interval. Use a short period if you are driving and a longer period if you are walking. Longer intervals use less battery power, and you are covering less ground while walking.

D-PRS

The Icom D-Star radios can transmit your GPS or manually entered location and other related information to other D-Star radios over D-PRS which is Icom's digital version of APRS (automatic packet reporting system). Normally your position will also be displayed on the aprs.fi website.

Figure 21: My D-PRS beacon (received using Echo)

This picture shows the information that my IC-9700 sends to other stations. I captured the image of my beacon by transmitting in Echo mode. Click on the red compass icon to the right of the scrolling received message to display any received station's position information. Provided they send it, of course. Note that the message that is sent to other D-Star radios is set in **Menu > Set > My Station > TX Message (DV).** It is not the same as the message that is sent to the APRS.fi website via the IP gateway. The remaining settings are in the **Menu > GPS > GPS TX Mode** area.

Note that the distance shown here is 11 metres from the location of the hotspot, that I entered into the DV repeater memory. If you were using a repeater, it would show the distance to that. The actual distance is only about 1 metre. I probably didn't enter enough decimal places.

D-PRS RADIO CONFIGURATION

There are many D-PRS and GPS settings. They are covered in detail in the Icom Advanced manuals, so I will only cover the basics here.

Turn on GPS

IC-9700: **Menu > GPS > GPS Set > External GPS** or **Manual**

IC-705, ID-51, and ID-52: **Menu > GPS > GPS Select** or **Manual Position**

GPS options

ID-52: The **GPS Option** setting has a sub-menu to select **SBAS** (space-based augmentation system) which is a set of geostationary satellites which are used to improve the accuracy of the GPS constellation. I cannot see any reason to turn this off. There is also an option to receive signals from the Russian GLONASS GNSS (global navigation satellite system). These complement the US GPS satellites. Again, I cannot see any reason to turn this off. **Satellite Information Out** sends the GPS data out of the USB port. You can select GPS data or GPS + QZSS + GLONASS data. These options are not available on the ID-51. **Power Save** turns the GPS receiver off for up to 8 minutes if the radio cannot see a GPS signal for five minutes. For example, you are in a cave... or inside the house. After that, it turns back on for another five minutes.

IC-705: **GPS Select** allows a choice of internal GPS, Off, or manual. **GPS Option** lets you add the SBAS and GLONASS GNSS satellites. The **Power Save** and **Satellite Information Out** options are the same as the ID-52.

The IC-9700 does not have a built-in GPS receiver, but you can connect an external receiver that has RS-232 levels. A cheap u-blox NEO or Adafruit GPS module with a TTL-RS232 level converter will also work. There is a baud rate setting for the receiver. They are usually either 4800 or 9600 bauds.

GPS TX Mode

Select **D-PRS**. You can select **NMEA** if you don't need to plot your position on the aprs.fi website, but D-PRS has more options.

Click on **D-PRS**. Leave the **Unproto address** in the default setting. These have been agreed with the APRS people and should not be changed. There is a table of the Unproto addresses for most models, in the technical information below.

Click on **TX Format**. There are four options. Position, Object, Item, and Weather. They each transmit a different set of information. The setup for each mode is quite similar. You can turn on or off items that you want to send or not send. The information sent by each setting varies from radio to radio. This is where you should look in the Advanced Manual for more information.

Position: is used when you are mobile or portable and you want to report where you are. 'Position' is the only mode that reports a live GPS position. All of the other options report a fixed manual position.

Object: is used to send more detailed messages. Perhaps some data about a SOTA activation or a DXpedition. The beacon contains a time stamp.

Item: is used to send more detailed messages. Perhaps some information about a repeater's location and modes. The beacon does not contain a time stamp.

Weather: is used for externally connected weather stations.

In most cases the best choice is **GPS TX Mode > D-PRS > TX Format > Position.**

Click on **Position.** Wow, even more choices!

Symbol: choose the one you like

SSID: the SSID is your callsign and a module character. The module character that you select should be the same as the one you use for your hotspot, or the one that matches the repeater you are using. (A for 23cm, B for 70 cm, or C for 2m). But you can make it anything from nothing to 9 and A to Z. Note that this letter will be sent on the position data for D-Star and for aprs.fi. I chose **-B** because I am working through a 70cm band hotspot. See the picture a few pages back.

Comment: This comment will be displayed on the aprs.fi website for all the world to see. Note that this is not the comment that other D-Star radios will see. The message that is sent to other D-Star radios is set in **Menu > Set > My Station > TX Message (DV).**

Time Stamp: The time stamp on the beacon transmission is in UTC. However, aprs.fi displays it in local time. You can choose from, HMS (hour minute second) or DHM (day hour minute) format. But aprs.fi displays it as, year, month, day.

Altitude: I leave it **off** because it is always inaccurate unless you are moving. Turn it on if you are on a mountaintop, mobile, or hiking.

Data Extension enables more choices. The display **course and speed** option is useful if you are in a car, boat, or aeroplane. **Power, height, gain, and directivity** can be used to convey information about your antenna system. It is fixed text fields, not live data.

NMEA mode

The NMEA mode passes the data strings from the internal or external GPS receiver straight through to the D-Star station you are working, and out of the USB port if you have that function turned on. It is more useful in that role because many mapping programs are designed to accept NMEA data. D-Star receivers will still display the familiar position screen, but your position will not be shown on aprs.fi.

You can send a message, the same as you can with D-PRS.

You can set the NMEA sentences to be transported. These carry different data from the satellites. Refer to the Icom Advanced Manual, my book about GPS, or an online search for details. The choices are RMC, GGA, GGL, VTG, GSA, and GSV.

FM APRS

The Icom radios do not support AX.25 APRS on FM. Only D-PRS on D-Star.

TECHNICAL INFORMATION

'D-PRS was developed to convert Icom GPS information generated by Icom D-STAR radios in GPS mode into TNC2 formatted APRS strings. It is important to note that D-PRS is not a protocol but a conversion specification. D-STAR digital voice mode (DV mode) is a 4800 bps digital data stream with no error detection or correction. Each transmission consists of a RF header followed by a 4800 bps stream of bits. 3600 bps of this stream is dedicated to the AMBE encoded voice. The remaining 1200 bps is used for synchronization and user-defined capabilities. Icom made use of this user-defined space to pass radio messages (simple display messages) and to pass serial data.'

'There is about 900 bps available for the serial data. Icom saw an opportunity to incorporate GPS information into this serial data. The intent was to maintain the primary purpose of DV mode, sending voice while providing a potentially valuable adjunct data stream, GPS information. We designed D-PRS to take this GPS information and make it presentable to the APRS world. D-PRS removes the need for designing special-purpose mapping software just for the unique Icom GPS data stream.

NOTE: It is very important that the UNPROTO does not contain any spaces and that it only contains the ",DSTAR" as the only "digipeater" in the path. The asterisk (*) following "DSTAR" is mandatory, as well.'* http://www.aprs-is.net/dprs.aspx

UNPROTO Settings GPS-A Mode	
Radio	UNPROTO
IC-2820	API282,DSTAR*
IC-705	API705,DSTAR*
IC-7100	API710,DSTAR*
IC-80	API80,DSTAR*
IC-9100	API910,DSTAR*
IC-92	API92,DSTAR*
IC-9700	API970,DSTAR*
ID-31	API31,DSTAR*
ID-4100	API410,DSTAR*
ID-51	API51,DSTAR*
ID-5100	API510,DSTAR*
ID-52	API52,DSTAR*
ID-880	API880,DSTAR*
	Table from http://www.aprs-is.net/dprs.aspx

APRS.FI WEBSITE

Most D-Star repeaters have D-PRS implemented on their gateway. If you turn D-PRS on and transmit through the repeater the D-PRS packet should eventually be displayed on the https://aprs.fi/ website. Go to APRS.fi, **enter your callsign in the search box**, and see your location plotted on the map.

This image is a beacon from a radio. It is different to the beacon from a hotspot.

If you are holding a normal conversation, your location should end up on aprs.fi. It sometimes takes a while, and it does not necessarily happen on every transmission. If you are testing or using the automatic beacon mode, you can unlink the reflector or gateway, or just use Echo, or the Local CQCQCQ mode, so that you are not repeatedly triggering every repeater that is connected to the linked reflector.

Note that if you selected NMEA instead of D-PRS on your radio, other D-Star radios will get your position information, but it will not be displayed on the https://aprs.fi/ website.

Automatic beacons

The data package sent to the APRS server is called a 'Beacon.' Using a repeater or reflector to send a beacon every few seconds is very annoying to other users. So please do not send automatic beacons more often than every 5 minutes. Turn off the automatic beacon mode if you are talking on the radio fairly often because your position will be sent every time you transmit anyway.

Sending your position to the aprs.fi website via a repeater.

There is no additional setup required. If your radio is capable of sending your D-PRS position to another D-Star radio, your beacon should eventually be relayed to aprs.fi and a beacon should be displayed on the map. It relies on at least one repeater that is transmitting your signal having an APRS gateway activated. If that includes your local repeater, you can transmit in Echo or local CQ mode and the beacon will still be sent to the aprs.fi website. If not, it will often work through a repeater that is connected to the same reflector. Of course, you would need to transmit in the Use Reflector CQCQCQ mode.

Sending your position to the aprs.fi website via a hotspot.

There is no additional setup required on the radio. If it is capable of sending your position to another D-Star radio, your beacon should eventually be relayed to aprs.fi and a beacon should be displayed on the map. You do need to turn on **Call Sign Routing** and **APRS Host Enable** and select an **APRS Host**.

PI-STAR HOTSPOT APRS / D-PRS

Pi-Star APRS beacon.

A Pi-Star hotspot can send an APRS beacon to aprs.fi completely independently. Even if you have D-PRS turned off in your radio. Turn on Call Sign Routing and APRS Host Enable on the 'General Configuration tab of the Pi-Star Configuration page. Select an APRS Host. There is a drop-down box choice of APRS gateway sites that will handle the data. The latitude and longitude location that is sent, is whatever you loaded into the Pi-Star configuration, not the GPS or manual location in the radio. ***Note that turning this on makes your location public.***

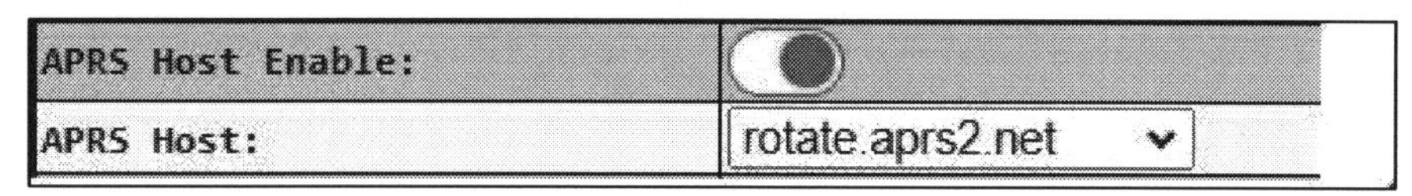

Figure 23: APRS setting in Pi-Star

Figure 22: Typical Pi-Star APRS beacon

The 'Pi-Star' beacon is sent when the Pi-Star boots after you power up, or if you click 'apply settings.' As far as I know, it does not send the beacon again.

TIP: There are several APRS Hosts to choose from. They are all "rotating" hosts. If an APRS server is 'down' or failed, the access will move to a different server.

Pi-Star Icom radio D-PRS beacon

The Pi-Star setting for passing D-PRS beacons from your radio through to the aprs.fi website is the same as above. Turn on Call Sign Routing and APRS Host Enable and select an APRS Host.

You do not have to be linked to a reflector. I use the Echo setting for testing. If you are linked to a reflector your position on aprs.fi will be updated when you transmit to other stations.

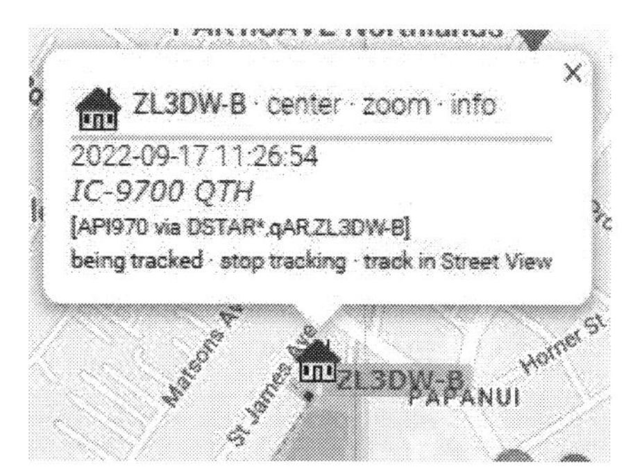

Figure 24: D-PRS beacon from the radio via the hotspot

Pi-Star Dashboard

This chapter covers the features of the Pi-Star dashboard after you have configured it for the hotspot and D-Star network. There is a lot of information presented on the main dashboard page. Click **Dashboard**.

Pi-Star Digital Voice Dashboard for ZL3DW

Dashboard | Admin | Configuration

Modes Enabled

Network Status

Radio Info

Trx	Listening
Tx	438.125000 MHz
Rx	433.125000 MHz
FW	MMDVM_HS:1.5.2
TCXO	14.7456 MHz

D-Star Repeater

RPT1	ZL3DW B
RPT2	ZL3DW G

D-Star Network

APRS	rotate.aprs2.net
	REF001 C DPlus/Out

Gateway Activity

Time (NZST)	Mode	Callsign	Target	Src	Dur(s)	Loss	BER
17:24:03 Sep 17th	D-Star	ZL3DW/INFO (GPS)	CQCQCQ	Net	7.1	0%	0.0%
16:55:05 Sep 17th	D-Star	ZL3DW/ECHO (GPS)	CQCQCQ	Net	1.1	0%	0.0%
16:55:02 Sep 17th	D-Star	ZL3DW/9700 (GPS)	E	RF	1.1	0%	0.0%
10:23:54 Sep 17th	D-Star	PY2NET/880H (GPS)	CQCQCQ	Net	1.2	[illegible]	0.0%
10:23:30 Sep 17th	D-Star	HL5BBD/7100 (GPS)	CQCQCQ	Net	1.2	[illegible]	0.0%
10:23:02 Sep 17th	D-Star	2E0KZV/ID51 (GPS)	CQCQCQ	Net	1.2	0%	0.0%
10:17:29 Sep 17th	D-Star	CT1AFS/DVSW (GPS)	CQCQCQ	Net	56.1	0%	0.0%
09:59:52 Sep 17th	D-Star	KA5BMC (GPS)	CQCQCQ	Net	0.4	0%	0.0%
09:58:25 Sep 17th	D-Star	KD9TSZ/DAN (GPS)	CQCQCQ	Net	0.3	0%	0.0%
09:46:38 Sep 17th	D-Star	G1LOE/5100 (GPS)	CQCQCQ	Net	0.7	0%	0.0%
09:42:02 Sep 17th	D-Star	KE8XQ/MARK (GPS)	CQCQCQ	Net	1.5	0%	0.0%
09:37:51 Sep 17th	D-Star	PA3AUM/EB0D (GPS)	CQCQCQ	Net	0.9	[illegible]	0.0%
09:33:02 Sep 17th	D-Star	N4OZI/ID51 (GPS)	CQCQCQ	Net	6.3	[illegible]	0.0%
09:32:52 Sep 17th	D-Star	W4CCD (GPS)	CQCQCQ	Net	0.5	0%	0.0%
09:21:05 Sep 17th	D-Star	KG7RNM (GPS)	CQCQCQ via REF001 C	Net	2.6	[illegible]	[illegible]
09:19:23 Sep 17th	D-Star	M7AGG/D74 (GPS)	CQCQCQ	Net	0.5	0%	0.0%
09:08:36 Sep 17th	D-Star	2E0ULH/8 (GPS)	CQCQCQ	Net	1.6	0%	0.0%
09:08:23 Sep 17th	D-Star	W5DKD P/ID31 (GPS)	CQCQCQ	Net	1.8	0%	0.0%
09:07:29 Sep 17th	D-Star	N7KB (GPS)	CQCQCQ	Net	1.0	0%	0.0%
09:07:22 Sep 17th	D-Star	W2HEF (GPS)	CQCQCQ via REF001 C	Net	1.0	[illegible]	0.0%

Local RF Activity

Time (NZST)	Mode	Callsign	Target	Src	Dur(s)	BER	RSSI
16:55:02 Sep 17th	D-Star	ZL3DW/9700 (GPS)	E	RF	1.1	0.0%	S9+46dB (-47 dBm)

Figure 25: Pi-Star Dashboard page

Modes Enabled shows the modes that you selected in the Configuration screen. This book only covers the D-Star selection, but you can have multiple modes selected.

Network Status. D-Star Net (green) indicates that the Internet connection is up, and the system is communicating with the D-Star network.

Radio Info shows the status of the hotspot.

Trx

- Listening (green) the hotspot is in its idle condition waiting for a call to arrive from the D-Star network or the RF path.
- Listening D-Star (grey) means that the user has stopped transmitting and the hotspot is waiting for the network hang time to expire. If no signal is heard during the hang time the hotspot will go to its idle condition.
- RX D-Star (dark green), means that the hotspot is receiving a signal from your transceiver.

TX and **RX** show the hotspot's transmit and receive frequencies, set in the configuration. They will be the same for a simplex hotspot and different for a duplex hotspot.

FW is the firmware revision of the hotspot.

TCXO is the nominal oscillator frequency. This is what you adjust if you apply an offset. Most hotspots run at either 12.288 MHz or 14.7456 MHz.

D-Star Repeater has information about the hotspot.

- RPT1 is the hotspot or repeater callsign and module code (B is 70cm, C is 2m)
- RPT2 is the hotspot or repeater gateway ID. It is the same callsign with a G module code

D-Star Network shows your network connection(s). Even if the APRS function is not turned on, APRS shows the APRS host that you selected. If you have 'Callsign Routing' enabled there will be a line showing the QuadNet link, ircv4.openquad.net. The final line shows the reflector that the hotspot is linked to, or 'Not linked.'

GATEWAY ACTIVITY

The gateway activity area shows calls broadcast by the hotspot that originated on the Internet (network) side. These are indicated with 'Net' in the Source column. Calls with 'RF' in the Source column originated as an RF signal received by the hotspot. Usually, that will only be your handheld or mobile radio(s). If the Pi-Star dashboard is for a repeater, there could be many RF calls.

Other information includes the date and time of the call, the mode will be D-Star unless other modes have been activated, the callsign that was received, and the target which is usually CQCQCQ unless a specific gateway or callsign was called, or E for Echo, or I for Information.

'Dur' is the duration of the call, Loss is the amount of data loss (after forward error correction) and BER is the bit error rate (quality of the signal). Clicking on the callsign takes you to the user's nominated website. Usually but not always their QRZ.com listing. Note that (GPS) does not mean that the radio is equipped with GPS. It is a link to the APRS.fi website with the callsign loaded as the search quantity.

LOCAL RF ACTIVITY

The local RF activity area provides pretty much the same information for calls that originated on the RF side of the hotspot. Plus, the received signal strength. For a hotspot on your desk, it is usually massive, S9 +46dB (-47 dBm). It is more useful for repeaters than hotspots.

THE ADMIN PAGE

The 'Admin' page is basically the same as the dashboard, with the ability to monitor the D-Star linking and link or unlink reflectors. You need to enter the username, **pi-star**, and password **raspberry**, the first time you access the admin page after you start the dashboard, or after the Raspberry Pi reboots. I recommend that you **do not** change the logon or password unless the hotspot is set for public access.

D-Star Link Information

Radio	Default	Auto	Timer	Link	Linked to	Mode	Direction	Last Change (NZST)
ZL3DW B	REF001 C	Auto	Never	Up	REF001 C	DPlus	Outgoing	17:23:54 Sep 17th

D-Star Link Manager

Radio Module	Reflector	Link / Un-Link	Action
ZL3DW B	REF001 C	Link UnLink	Request Change

The 'D-Star Link Information' is just copied from the configuration tab. **Radio** is the callsign of the hotspot. It is almost always your callsign unless Pi-Star is being used for a repeater or a public hotspot, in which case it would be the repeater's callsign. **Default** and **Auto** show the default reflector link that will be made after you reboot the hotspot. It can take up to a minute to connect. 'Manual' would indicate that the hotspot will not automatically connect to a reflector. **Timer** is normally set to 'Never.' It is linked to a setting which resets the hotspot every hour. **Link** should be 'Up.' As indicated by the green D-Star Net icon.

Linked to and **Mode** indicate the reflector and type of linking being used. **Direction Outgoing** means that you initiated the link, and the **Last Change** indicates when the link was established.

At the top of the page, there is **Gateway Hardware Information** about the Raspberry Pi. It shows the Host Name (which you should not change), The Pi Kernel release, The Pi model, its CPU loading per core, and the CPU temperature. If the temperature gets into 'the red' you need to add a heatsink or fan to improve the Raspberry Pi CPU cooling. Mine sometimes runs in the orange region at 53.5 degrees Celsius, but it seems to be stable. I do have a heatsink on the CPU.

I don't know much about the **Service Status** section. I guess, "green is good." MMDVMHost means the hotspot is working as an MMDVM host. ircDDBGateway (IRC-based Distributed Database) is the international D-Star database. It handles callsign and routing information between gateways. If you have not registered your callsign the system will not know where your hotspot or repeater is connected and will be unable to route traffic to you. The other green icon is PiStar-Watchdog which indicates the status of the Pi-Star software. If there is a problem the timer will elapse, and the indicator might go red. It is more likely that the hotspot will just freeze up.

Even though it is the "Admin" page, the reflector linking is the only setting you can change. But you do get three more menu items. **Live Logs** (records incoming and outgoing calls), **Power** to shut down or reboot the Pi, and **Update**. Only click this if you want to do an update. It takes several minutes, and you won't get asked again!

THE CONFIGURATION PAGE

If you are configuring a Pi-Star hotspot from scratch, check out the 'Hotspots,' 'Loading Pi-Star,' and 'Pi-Star Configuration' chapters. This section covers the configuration of the hotspot after you have it working and connected to your local network.

The configuration page adds some more menu items at the top of the page. **Expert** for experts like us, **Power** to shut down or reboot the Pi, **Update**, only click this if you want to do an update. You won't get asked again! **Backup/Restore**, and **Factory Reset**. Do not click that or you will erase all your configuration settings. Unlike the DMR setup, you probably will not have to access any of the Expert pages. It is better to leave them alone unless you really are an expert.

Backup/Restore

You should make a backup as soon as your configuration is stable and working. Backups are downloaded to your standard Windows download directory as a Zip file. You can leave them there or move them to another folder.

The zip file contains twenty configuration files. Editing settings in the 'Expert' area changes the relevant config file. Most are text files that can be viewed or edited if you are especially brave, in Notepad or Wordpad. I have copied text from an old file to correct a DMR network configuration error after I messed around with the settings.

You can restore the configuration from the Zip backup file if you want to revert to an older configuration or copy your hotspot configuration to a new hotspot.

Apply changes

You will see **Apply Changes** at the bottom of each section. It should be used every time you make changes in any of the configuration sections. New options often appear on the configuration forms after you click **Apply Changes**. Click the button and wait for the hotspot to reboot, then check what changed before proceeding to the next section.

Control Software section

Select **MMDVM Host** for the 'Control Software' option. The repeater option is for a repeater controller card, so you probably won't use it.

'Controller Mode' should be set to **Simplex Node** for a simplex hotspot, or **Duplex Repeater** for a duplex hotspot. I use a duplex hotspot.

TIP: There is no big advantage in using a duplex hotspot for D-Star. There ***is*** *a big advantage in using a duplex hotspot for DMR. It allows you to transmit on one time slot while receiving a signal on the other, or to receive two timeslots at the same time. A simplex hotspot will block you from transmitting if a call is being received on the other time slot. D-Star repeaters cannot transmit two voice signals at the same time, so a simplex hotspot works just as well as a duplex one. If you plan to use DMR sometime in the future, buy a duplex hotspot.*

MMDVM Host Configuration section

This is where you select which mode the hotspot will work on. Select D-Star Mode by clicking the box beside the label.

You can select other modes such as YSF or DMR if you have radios for those modes.

Leave the hangtime settings at the default value of 20. It can be confusing when calls come in and you could miss part of a conversation on one mode if the hotspot switches to a different mode. The hang time settings create time (20ms) for someone to come back to your call before a call on another mode can take over. If you are using multiple modes and this becomes a problem, you can increase the hangtime to alleviate the issue.

The MMDVM Display Type is set according to what display (if any) you have on your hotspot. See MMDVM Display on page 138.

If you have made any changes, click **Apply Changes**.

General configuration

The general configuration section contains information about the hotspot and your location. It remains the same no matter what mode you select, so it should already be configured. The hostname should be pi-star, I don't recommend changing it. The node callsign will usually be your callsign unless the Pi-Star configuration is for a public access repeater. If you select **Auto**, on the URL line the software will choose your QRZ listing address. You can select **Manual** and enter a different website if you want to. Again, this option is really for public access repeaters. Radio/modem type is critical. It must match your hotspot. But it should have been set up by now. See the Hotspots and Pi-Star chapters for setup details.

The Node Type should be **private** unless you have a good reason to make it public.

Only turn on **APRS Host Enable** if you want to send your position to the aprs.fi website. This makes your location public. The hotspot will send a beacon to aprs.fi whenever it is re-booted, and the radio will send a beacon to aprs.fi on every transmission. If you leave it turned off, the D-Star stations you work will get the usual D-PRS position display but there will be no beacons sent to the aprs.fi website.

The APRS Host dropdown box provides a choice of 'rotating' APRS host sites. Choose one near you.

The time zone and language were set in the initial setup. Usually as a part of the Pi-Star download from pi-star.uk.

If you have made any changes, click **Apply Changes**.

D-Star configuration

The D-Star configuration section will be shown once you have selected the D-Star mode on the MMDVM Host Configuration and clicked **Apply Changes**.

RPT1 Callsign is the callsign of the hotspot (or repeater) node. The callsign was set in the General Configuration section, but you can add a module code. Normally the module code relates to the frequency of the hotspot or repeater. A = 23cm, B = 70cm, or C = 2m. But if you already have a hotspot using that module code, you can select any other code except G. You cannot have two hotspots using the same module code.

RPT2 Callsign is the gateway callsign of the hotspot (or repeater) node. It is always set to the node callsign with a G module code.

The **Remote Password** is the **raspberry** password required to access the Admin and Configuration pages. Don't change it unless you are making the hotspot public access. If you are making hotspot public access, you should definitely change the remote password to stop Internet users from changing your hotspot or repeater configuration.

The **Default Reflector** is the reflector that the hotspot will connect to when it is booted or restarted with the Apply Changes button. Set it to the reflector that you use the most. Select **Startup**. If you select Manual, the hotspot will not connect to a reflector on startup. It will stay 'Unlinked' and you will have to establish a link manually.

The **ircDDBGateway** language defaults to English but there are many other settings. **Time announcements** enables hourly time announcements via the radio. It might be handy if you have a sked, or like me, you need reminding when to go to bed.

Callsign routing establishes a connection via the ircDDBGateway code in Pi-Star. Theoretically, it allows routing calls including Gateway CQ routing and Icom Call Sign Routing to an individual callsign by entering a person's callsign into the UR field using Direct Input (UR). And it should allow you to call 'Smart Routing Groups,' often referred to as' QuadNet smart groups' or 'STARnet groups.' I have not been able to get any of these options to work.

If you have 'Callsign Routing' enabled there will be a line showing the QuadNet link, ircv4.openquad.net on the D-Star Network section at the bottom left of the Pi-Star dashboard. If it says rr.openquad.net the Pi-Star needs an **Update**, or you can change the text to ircv4.openquad.net in the Expert settings.

Use DPlus for XRF, makes connections to XRF reflectors using the DPlus protocol, the same as REF reflectors rather than using the DExtra protocol. All XRF reflectors accept either protocol. Some home networks don't play well with DExtra because it requires port forwarding of port 30001 in your Internet router. Enable this function if you find that you are unable to link to XRF reflectors. If you change this, you have to do a full Pi-Star **Update** as well. Not just a re-boot.

Mobile GPS Configuration

This setting has nothing at all to do with the GPS receiver on the radio. It is used when you connect a GPS receiver dongle to the Raspberry Pi that is hosting the MMDVM hotspot. You can turn on the serial port or USB port if a USB dongle is plugged in and the baud rate. Normally 4800 or 9600 for a GPS receiver but may be faster for a dongle.

Firewall configuration

Firewall configuration is for experts. Leave all the access settings set to **Private** and Auto AP **on** and uPNP **on**. Auto AP lets the Raspberry Pi act as a WiFi access point if it is unable to find your WiFi network within 120 seconds. You can link your phone to the node and configure the WiFi access on the Raspberry Pi. uPNP lets the Raspberry Pi manage its own firewall settings. If you turn it off, you can configure the firewall settings in the Expert tab. Setting Dashboard access to **public** would allow Internet users to observe the Pi-Star dashboard. This is OK providing you changed the 'Remote Access Password' to stop them from changing the configuration settings. Making the ircDDBGateway public would allow others to manage routing. Changing SSH access enables remote users to access the Raspberry Pi via Secure Shell. This could be required if someone is offering remote support to fix a Raspberry Pi configuration issue. Note that SSH access within your LAN is always available. You can access the Raspberry Pi via the puTTY SSH (secure shell) program without changing this setting.

Wireless Information and Statistics

Wireless Information and Statistics shows the WiFi connection. It should state 'Interface is Up.' You can configure, reset, or refresh the WiFi if it is not working.

Auto AP SSID

This setting changes the default so that a password and login name is required to access the hotspot in AP mode. If you need the AP mode, you are probably in enough trouble without adding this extra layer of complexity.

Remote Access Password

The remote access password should only be set if you are making the hotspot public. You would normally only do that if you were using Pi-Star on a repeater, and you wanted to make the dashboard available for repeater users.

If you are the only person that needs to see the hotspot dashboard, keep it set to private, and do not enter a remote access password.

POWER

The **power** menu item provides this attractive image. Click the green **reboot** icon to restart the hotspot and click the red **shutdown** icon to safely shut down the hotspot. The file system on the Raspberry Pi SD card can become corrupted if you just pull the power out. It is always best to shut down the Pi using a button on the hotspot display if it has one (mine does not), or this power button. I admit that I usually just pull the power plug, but that is a risk that I take, and I can restore the software if I have to.

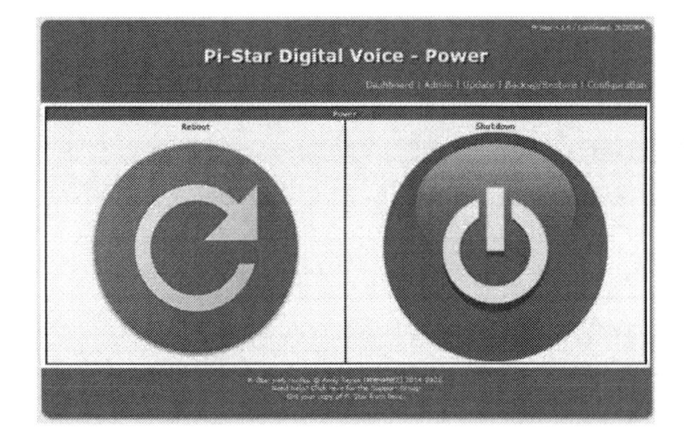

Hotspots

If you have a local D-Star repeater you do not need a hotspot. However, it is nice to be able to connect to any reflector or gateway without having to worry if someone else was using the link you just closed.

Buying hotspots is a bit addictive. The next thing is, you will want two, so you can monitor two reflectors at the same time, or to support a DMR or YSF radio.

Most of the hotspots "out there" are MMDVM simplex hotspots. They transmit and receive on the same frequency. Most models can be used for some, or all, of the digital voice modes, D-Star, DMR, Yaesu System Fusion (C4FM), NXDN, POCSAG, and P25. I am only going to talk about D-Star. Most MMDVM hotspots use a Raspberry Pi Zero W or the Pi Zero 2W single-board computer to run the Pi-Star routing software. All of those hotspots can also run on a Raspberry Pi 3, or Pi 3 Model B, but the larger form factor does not match the size of the hotspot. Duplex hotspots are bigger and need more computing power, so a Raspberry Pi 3B is preferred. The Pi 3B supports 2.4 GHz and 5 GHz WiFi, but the Pi Zero W only supports 2.4 GHz WiFi. Raspberry Pi Zero boards are much slower, particularly when booting up, but the gap is narrower with the new Pi Zero 2 which is five times as fast as the old Pi Zero. Some hotspots use other single-board computers. Duplex hotspots have two antennas and they run in duplex mode like a repeater.

Note that the Pi3 setup requires a bigger power supply than a Pi Zero with an OLED screen. The recommended supply is 2.5A (700mA minimum). I have had no problems running my hotspot off a powered USB 3 hub. The 2.4" Nextion screen draws 90 mA. You need an additional USB to TTL adapter to use Nextion screens with most simplex hotspots, but the Rpi 3B serial port can interface with the screen directly.

TIP: You don't need very much transmitter power to reach a hotspot in the same room as your radio. The Icom ID-52 has a very low power mode at 0.1 watts (100mW) which is ideal for hotspot operation.

MMDVM RASPBERRY PI HOTSPOTS

There are dozens of different models which may come with plastic cases or Perspex sheet protection. Most dealers require you to buy a Raspberry Pi Zero or model 3B elsewhere. A search online will bring up many options. Here are a few I found online. Note I am not endorsing any model. I have not tried most of them.

TIP: It is often difficult to work out what you are going to get. Especially with the Chinese suppliers. For example, some of the websites show a picture of a hotspot with a Pi Zero. But they do not ship with a Pi Zero. The rule is. If the advert does not specifically say something is included… it is probably NOT included. Even if there is a photo showing the item. Some hotspots are supplied assembled, some are not. Some include a case, and some don't.

BI7JTA duplex hotspot. Unfortunately, the model I bought is no longer available. But similar models are available online. I bought a duplex hotspot that was provided with a 2.4" Nextion screen, the programmed SD card, and the Raspberry Pi model 3B. https://www.bi7jta.org/cart/.

TGIF Spot. https://tgifspot.com/ has a range of simplex hotspots. One with an OLED display, one with a 2.4" Nextion display, and one with a 3.5" Nextion display.

RFinder created the SkyBridge+ dual-band simplex and the HCP-1 duplex hotspot. The HCP-1 includes an internal battery and is very portable. The computer is a Raspberry Pi Zero.

LZ duplex and LZ simplex hotspots, some come with a 3.2 Inch Colour Screen. They are MMDVM hotspots designed for a Rpi Zero W.

Jumbospot dual-band simplex MMDVM - Rpi Zero or 3B. These are a clone of the ZUMspot. Sold from a wide variety of vendors with a wide variety of prices depending on what you get. Most often they do not include the Raspberry Pi Zero, and sometimes they do not include the case.

ZUMspot https://www.hamradio.com/detail.cfm?pid=H0-016491. These are simplex, UHF hotspots with a 1.3" OLED screen. They require you to buy a Raspberry Pi Zero W or Zero 2W. They will also run on an RPi 3B or an Odroid SBC.

Rugged Spot supplies a series of 'NEX-GEN' simplex MMDVM hotspots, some with Nextion screens and some with ceramic antennas (which work very well). They are supplied pre-programmed which is very nice. They are based around a JumboSpot supplied with a Raspberry Pi3-B and a plastic layer case.
https://hamradio1.com/product/rugged-spot-store/

TIP: Don't worry if the hotspot is not assembled. All you will have to do is solder the RPi header pins and sometimes the RF SMA connector (make sure those are supplied though). My simplex hotspot kit came with all the required screws and two short headers for the Raspberry Pi. The hotspot board was 100% complete. You will need an SD card and some free software to create the Pi-Star image and probably a micro USB cable or a USB power supply, to power the hotspot.

SHARKRF OPENSPOT 4 HOTSPOT

The SharkRF OpenSpot 4 has a built-in battery. It is a fully cased simplex hotspot with WiFi. Instead of Pi-Star, you use 'SharkRF link' for configuration using any web browser. Configuration and connection to a WiFi network seem to be very easy. They are more expensive than the MMDVM simplex hotspots. If you intend to use one for DMR, note that the OpenSpot 4 can only connect to one DMR network at a time. The SharkRF OpenSpot 4 Pro version supports D-Star crossover modes.

You can use a D-Star transceiver to access DMR, YSF, or NXDN networks, or your DMR, YSF, or NXDN transceiver to access D-Star networks.

SIMPLEX HOTSPOTS

Most people use simplex hotspots. They are cheaper, easier to set up, and faster to use than duplex hotspots. Simplex hotspots receive and transmit on the same frequency. Almost all simplex hotspots are MMDVM modem boards paired with Raspberry Pi Zero W or Zero 2 W single-board computers. I prefer models with an OLED display. It is easier to tell whether they are working.

Raspberry Pi Zero

You will probably have to buy a Raspberry Pi Zero to supply the computing power for your MMVM hotspot. Very few hotspots are supplied with one. You must get the 'Zero W' version which has WiFi. I recommend getting the new Raspberry Pi Zero 2W. It is five times faster than the old Zero W and it only costs around $15 US, (about £13 UK).

The SD card and display

If you are providing your own Raspberry Pi, you will also have to buy and prepare a micro SD card for the Raspberry Pi. It contains the Linux operating system and the Pi-Star modem software for the hotspot. I cover that in the 'Loading Pi-Star' chapter starting on page 126.

Because of the small form factor of the MMDVM modem board and the Raspberry Pi Zero, most simplex hotspots have a small 1.3" OLED display or no display at all. But it is possible to connect a Nextion display to the Raspberry Pi on a simplex hotspot using a USB to TTL adapter.

TIP: The Nextion displays work better with a duplex hotspot on a Raspberry Pi 3B.

Building a simplex hotspot

Don't be scared of building a simplex hotspot. It is just an assembly job. The MMDVM modem board will be supplied 100% complete. You might have to solder some header pins and possibly the SMA connector. No problem if you have a temperature-controlled soldering iron and at least some experience with soldering components onto a printed circuit board. Usually, there is a more expensive option to buy a board where no soldering is required.

You may be required to solder in some header pins on the Raspberry Pi Zero. Again, there is usually an option to pay more for one that has the header pins soldered.

The soldering is not difficult. Give it a go!

My hotspot kit included the case, screws, spacers, and even the header pins for the Raspberry Pi. The picture shows the kit that I received. The pcb was supplied in an anti-static bag and no soldering was required on the hotspot board. Just plug the hotspot board onto the Rpi and install it into the case. The long spacers go on the side that is not supported by the header pins.

Figure 26: A typical simplex hotspot kit. No soldering required.

The case clips together and it is a bit tricky to open. I wedged one side open with a small flat screwdriver and slipped a strip of cardboard into the gap. Then I applied the screwdriver to the other side and the case slid apart easily. Don't reassemble the case until the board has been mated with the RPi Zero and screwed onto the base section. And remember to remove the protective plastic from the OLED screen by pulling the small tab.

Figure 27: Raspberry Pi Zero 2W (no header pins)

Solder the header pins into the rows at both ends of the board. Solder the short pins to the PCB. The longer pins should face up.

I chose to pair the MMDVM modem with the Raspberry Pi Zero 2W. They usually have no header pins installed, but they were supplied in the hotspot kit. If the Raspberry Pi Zero has the full header row you can throw away the supplied header pins or keep them for another project.

Put both sets of pins into the board before you flip the board over so that it sits flat when it is upside down for soldering. The plastic must lie flat on the board so that the pins are exactly vertical.

Figure 28: Raspberry Pi Zero W (with header pins)

My Pi Zero was supplied with header pins.

Figure 29: Snip these four leads short if the RPi has a full row of header pins

WARNING: If the Raspberry Pi has the full row of header pins, use side cutters to trim the four pins on the display board so they won't short out pins on the Raspberry Pi GPIO header.

Place the Raspberry Pi into the case making sure that the SD card slot is accessible. Insert the two short screws on the header side of the Raspberry pi board. Carefully stack the boards, using the supplied spacers on the side that is furthest from the header plugs. Insert the long screws through the hotspot modem board, the spacers and the Raspberry Pi and tighten, (not too tight).

Clip the case on being careful not to damage the micro USB and HDMI jacks on the side. Ease the case around them and they will pop into the shell. Screw the antenna onto the SMA connector and power up the board.

IMPORTANT: The micro USB closest to the end is the power connector.

Notes for a simplex Pi-Star hotspot

The Pi-Star configuration is covered in the Pi-Star configuration chapter. There are some minor differences when using a simplex hotspot. Click **Apply Changes** regularly after each configuration step. It takes some time for the modem to reboot, but the configuration screen may change as a result of your previous choices. If something looks odd, such as two entry boxes for frequencies, click **Apply Changes** and the page will reformat.

- The controller mode is Simplex
- Only one frequency
- The display option is usually OLED /dev/tty/AMA0
- The Radio/Modem type is different to a duplex hotspot

Figure 30: My MMDVM simplex hotspot

DUPLEX HOTSPOTS

More and more people are using duplex hotspots, to take advantage of the bigger display and the two timeslots offered by DMR. There is no big advantage in using a duplex hotspot for D-Star and other digital voice technologies. Duplex hotspots are larger and more expensive. They will work with a Raspberry Pi Zero, but the form factor and computing requirements make a Raspberry Pi 3 Model B a much better option.

TIP: If you think that you would like to try DMR at some stage, buy a duplex hotspot. Otherwise, a simplex hotspot is just as good.

Duplex hotspots act like a repeater, transmitting and receiving simultaneously. If you have two radios on the hotspot frequency and you transmit, the other will receive the signal back from the hotspot. They have two RF chips and two antennas, and they use two frequencies. You can set any frequency split between receive and transmit, but I believe it is best to stick with the standard ±5 MHz offset on the 70cm band and ±600 kHz offset on the 2m band.

Bigger display

My duplex hotspot is driving a Nextion display giving a nice bright colour display of the D-Star activity. You can see a Canadian caller on REF001 C. The hotspot is running on a Raspberry Pi Model 3B and uses the Pi-Star software. Note that the frequency information is embedded in the Nextion background image and is usually wrong.

Figure 31: A duplex hotspot receiving VE4SET on REF001 C

POWERING UP YOUR HOTSPOT

Make sure the antenna or antennas are plugged in. Some models have a ceramic antenna on the board, but most simplex hotspots have a small SMA antenna. Duplex hotspots usually have two SMA antennas. I have not seen one with ceramic antennas.

Plug the power cable into the Raspberry Pi. It uses a micro USB connector. You can power a Rpi Zero from a plug pack, a USB 3.0 port on your computer, a USB hub, or a USB power supply. A Raspberry Pi 3B is a bit more power-hungry than a Pi zero, especially if you have a Nextion display. Many phone chargers and USB2.0 ports cannot supply enough power and the device may not work reliably. The official Raspberry Pi power supply is capable of supplying 2.5 amps at 5.1 volts. You may also need a 'micro USB' to 'USB type A' cable.

A Raspberry Pi 3B will boot quite quickly. If you have a Nextion display it will show the idle screen as soon as the power is applied. The screen pages are coded into the display, not the Raspberry Pi program. It will take a minute or so for the WiFi to get established with a DHCP network address. If you have a Nextion screen the IP address might be listed at the bottom right of the screen. Eth0 means a wired Ethernet Connection and wlan0 indicates a WiFi connection.

The Raspberry Pi Zero will take a minute to boot, followed by some more time for the WiFi to get established with a DHCP network address.

TIP: If it is the first boot and you copied the wpa_supplicant.conf into the SD card boot drive, the Rpi will boot, load the WiFi config, and then reboot. It could take several minutes.

MY NEW HOTSPOT DOESN'T WORK

Don't panic! I have covered this situation in the troubleshooting section, on page 156. The most common problems are,

- An insufficient power supply. Symptoms, include Windows 'bonging' regularly, hotspot rebooting, no display, or incorrect display.
- The display is not configured correctly in Pi-Star. No display even when rebooting the hotspot.
- Incorrect Modem setting in Pi-Star. No transmission to the radio. No display.
- Hotspot being well off frequency. Seeing activity but no audio from the radio, no display of calling stations, hotspot not responding when you transmit.
- Network disabled in Pi-Star. Check the bottom right box on the Pi-Star dashboard.

PI-STAR TECHNICAL INFORMATION

Almost all hotspots use the Pi-Star modem software. Its function is to manage the routing of data traffic received by the hotspot board to the D-Star network. It also routes the traffic from the D-Star network to the hotspot board which transmits it to your handheld or mobile radio.

Pi-Star is extremely capable and, if you get into the 'Expert Level' functions, very complicated. We will start with a basic setup and proceed from there with caution. I only have a very superficial understanding of Pi-Star, "just enough to get me into trouble." This quote from the website sums it up nicely.

"The design concept is simple. Provide the complex services and configuration for Digital Voice on Amateur Radio in a way that makes it easily accessible to anyone just starting out but makes it configurable enough to be interesting for those of us who can't help but tinker." [Pi-Star UK].

Pi-Star runs on the Raspberry Pi. The dashboard interface is just a web page that can be accessed using your favourite Internet web browser. You make changes on the web page and save them back to the hotspot. You can access the Pi-Star dashboard from your PC, tablet, iPad, or even a phone. Normally for a home hotspot, your controlling device has to be connected to your local LAN at home. However, you can configure Pi-Star for public access over the Internet and you sometimes see this with publicly accessible repeaters and high-power hotspots.

Pi-Star includes the Raspberry Pi OS Linux distribution from the Raspberry Pi Foundation. It is a variation of Debian Linux that has been optimised to run on Raspberry Pi ARM boards. Check out the 'What is Pi-Star' page on the Pi-Star.uk website.

THE PI-STAR WEBSITE

The Pi-Star website is at https://www.pistar.uk/. Note that if you have a Pi-Star hotspot plugged in and you type 'Pi-Star' into the URL box, you are likely to end up on your Pi-Star dashboard instead of the Pi-Star website. Pi-Star is based on the DStarRepeater and ircDDBGateway software designed by Jonathan Naylor G4KLX, which has been extended to support the full G4KLX MMDVM suite, including the extra cross-mode gateways added by José (Andy) Uribe, CA6JAU.

The website has a link to the D-Star registration page and downloads for the latest revision of Pi-Star for various hardware platforms.

HOTSPOT SD CARD

Unless your hotspot came with a fully programmed SD card, you will have to buy one. You can fit Pi-Star onto a 2 Gb micro SD HC card, but you probably will not be able to find one that small. I ended up with a 32 Gb card and 27 Gb of free space.

The Raspberry Pi is not tremendously fast, so there is no need to buy a superfast SD card. The card will probably already be formatted, but if not, you can format it with the FAT32 option. The image will overwrite the formatting with a Linux format anyway. SD cards are available from most electronics and computer shops, or the usual online sources. I have had good results using SanDisk cards, but any micro SD HC card should be fine. Kingston and Samsung disks are also recommended. I don't like Adata products. I had a bad experience with one of their SSDs.

TIP: If the card has already been formatted for Linux, Windows will flip out and try to open each partition. It will also ask you to format each partition. Just keep clicking the close button on all the popup windows until it quits asking. This is a real pain, and it will happen any time you put a Linux SD card into a Windows machine. If you happen to have a Linux machine you can bypass this problem by flashing the SD card on that.

Figure 32: A micro SD HC card and free adapter

SD card reader & writer

If you have a notebook PC, it may have a built-in SD card reader. If it does, it will probably be for the full-size SD cards, but you can buy a micro SD HC card that comes bundled with an adapter. If like me, you use a desktop PC without an SD card reader, you can buy a USB SD card reader. They write as well. They only cost $5 US or £4 or thereabouts. You can use one that takes full-size SD cards and buy a micro SD card that is packaged with an adapter, or you can buy one that takes the micro SD card. I have one of each. One of them came free with a unit I bought from China. USB card readers are available from electronics and computer shops, and the usual online sources.

Downloading the Pi-Star image

Unless your hotspot came supplied with a pre-programmed SD card for the Raspberry Pi, the Pi-Star software must be downloaded from the Pi-Star website.

This is a big 612 Mb download. It will require a good Internet connection and it can take a long time, depending on your broadband speeds. The download took about 4 minutes at my place.

Open the Pi-Star.uk website in your favourite web browser and select **Downloads > Download Pi-Star** from the menu bar on the left. Or go straight to https://www.pistar.uk/downloads/. There are download options for several single-board computers. NanoPi Air, Nano Pi Neo, Odroid XU4, Orange Pi Zero, and the DV Mega dongle. There may be several releases for the Raspberry Pi listed.

The RPi versions are OK for the Pi Zero and the Pi 3B. Download the newest release by clicking the orange text. The current release is **Pi-Star_RPi_V4.1.5_30-Oct-2021.zip.**

When the zip has finished downloading, unzip the files to a directory. Perhaps a Pi-Star directory or a Temp directory. It does not matter so long as you can find the files again. There are two files, a .img file which is the Pi-Star image and a .md5sum which is a checksum file used to detect whether the image file is corrupted.

Flashing the SD card

The website has a good set of instruction guides for flashing the image to an SD card, but I am going to step you through the process anyway. There is also a video at https://www.youtube.com/watch?v=B5G4gYDdJeQ.

There are a few programs that can write an image file to an SD card. I have tried 'Win32 disk imager' and 'Balena Etcher.' They both work well. I usually use Balener Etcher for Linux SD disks. If you don't already have it, download one or the other and install it on your PC.

"Send in the clones." If you have a friend with a working hotspot you might like to make a clone of their image file rather than download the latest version from Pi-Star.uk. You can use either program to make a clone of a working SD card onto a new SD card. Note that this is not the same as making a copy of the files. It is copying the file structure and formatting as well. Technically it should be possible to format the card on a Linux machine and then copy the files, but I won't guarantee that won't end up causing you hours of tinkering.

Insert the SD card into your PC, either directly or in the USB-to-SD adapter. As discussed above you may have to use a micro SD to SD adapter as well. These come free with many micro SD cards. Close any annoying error popup windows from Windows. Do not follow the advice to format the partitions. **Make 100% sure that you know the drive letter of the SD card. You do not want to write a Linux distribution onto a USB drive you happen to have plugged into your PC.**

Balena Etcher method

I like Balena Etcher because it is straightforward and harder to make a mistake. Click Flash from file and navigate to your .img file.

Click Select Target. It should find your SD card automatically but check the drive letter to make sure.

Click Flash. Wait until the write and validation cycle is complete. Close the Windows error windows, do not format the disk. Remove the SD disk and the USB adapter from the USB port.

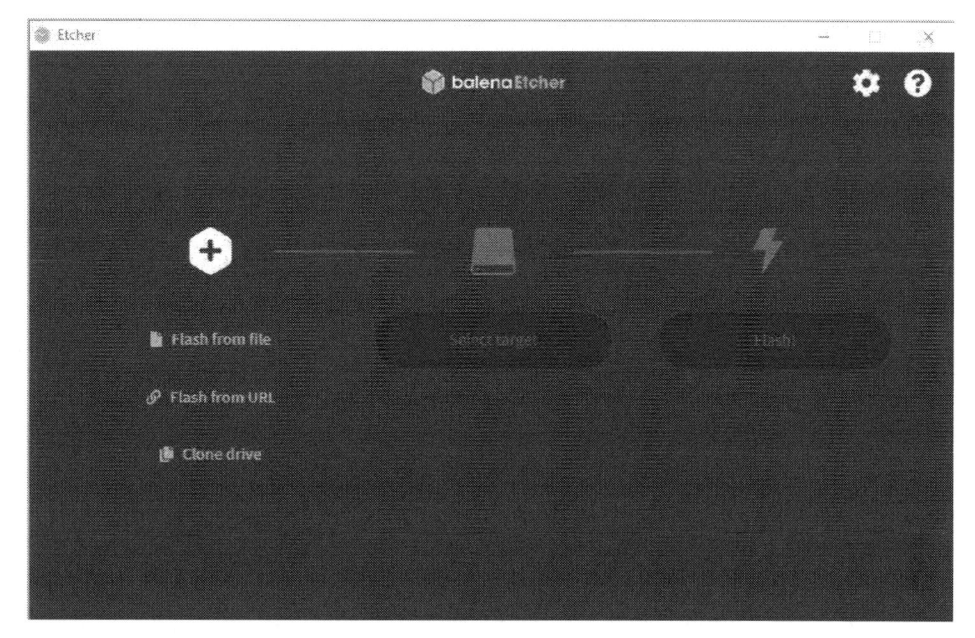

Figure 34: Balena Etcher

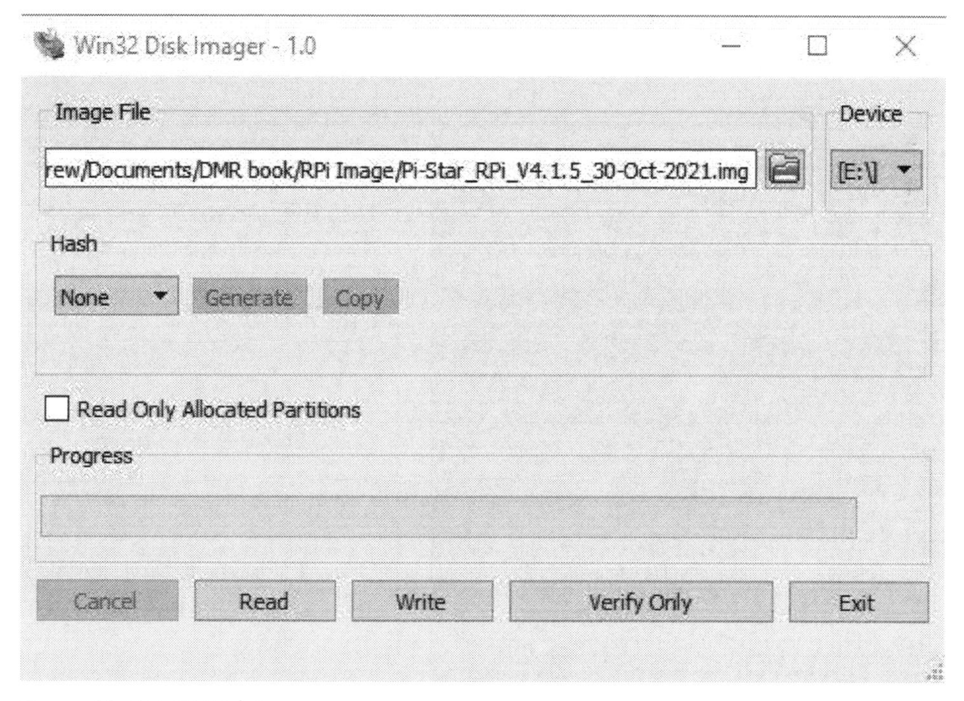

Figure 33: Win32 Disk Imager

Select target 4 found

	Name	Size	Location		
☐	⚠ WDC WD1003FZEX-00MK2A0	1 TB	G:\	Large drive	Source drive
☐	⚠ Samsung PSSD T7 SCSI Disk Device	1 TB	E:\	Large drive	
☑	Mass Storage Device USB Device	32 GB	F:\		

Figure 35: Balena Etcher - choose the SD card

Win32 Disk Imager method

Click the blue disk folder icon to the right of the top 'Image File' text field.

Use the 'Device' dropdown to select the SD card. Check the drive letter to make sure that it is the SD card and not some other drive.

Ignore all the other settings and click the **Write** button. The writing process will be shown on the progress indicator, then the validation process. It can be quite slow.

Wait until the write and verify is complete. Close or Cancel the Windows errors. Do not format the disk. That would overwrite what you have just placed onto the disk. Remove the SD disk and the USB adapter from the USB port.

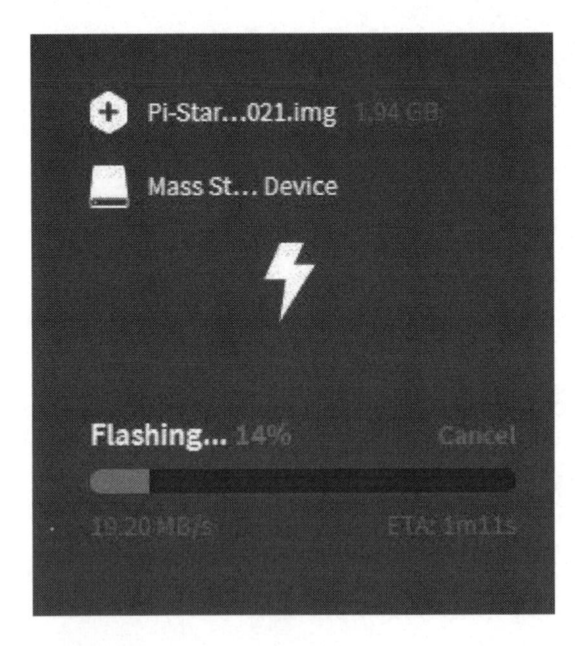

Select the SD card. Note the warnings on the large drives. Check that the 'Size' matches the SD card and note the drive letter.

Click Flash and wait for the disk write and validation to complete.

Cancel any Windows errors and do not format the new disk volumes.

LAN CONNECTION

Duplex hotspots running on Raspberry Pi 3 or 3B boards can be connected to your LAN with a direct Ethernet connection to your main router, or a hub. I have a four-port internet switch in the shack for connecting my HF transceivers and my PC. However, most people configure the hotspot for WiFi access, and I ended up using a WiFi connection as well. My duplex hotspot helpfully shows the IP address in the bottom right of the idle screen when there has been no talk group activity for a while.

Simplex hotspots always use WiFi because the Raspberry Pi Zero does not have an Ethernet connector.

WIFI CONNECTION

The SD card is complete, but the Raspberry Pi will not have access to your WiFi. Before you proceed you need to know the name of your WiFi network, which is referred to as the SSID (service set identifier) and the WiFi password which is called the PSK (pre-shared key).

TIP: This is the same information that you would use if you were adding a phone or computer to your WiFi network. If you don't know the WiFi name and password, it may be on the back or underside of your WIFI Internet router. Or in the booklet that came with it.

If you are using a Raspberry Pi Zero, you need to complete the steps in configuration option 1 or option 2.

If you are using a Raspberry Pi 3 or 3B, the easiest method is to use configuration option 3. Which involves using the wired Ethernet connection to configure the WiFi connection. But you can use option 1 or option 2 if you want to. If you plan to always use a wired Ethernet connection with your Raspberry Pi 3 or 3B, you don't need to set up a WiFi connection at all.

Option 1: Adding the WiFi settings using the Auto AP method

For this method, you need a computer, phone, or tablet with a WiFi connection. It avoids having to download a 'wpa_supplicant.conf' file and copy it into the boot directory of your new SD card. It is also useful if you take your hotspot somewhere and want to connect it to a different WiFi network.

Pi-Star can create its own WiFi network. When you power up the hotspot it will take a while to boot up. The Pi Zero could take up to a minute. If the hotspot is unable to connect to a WiFi access point within two minutes after it finishes booting up, it will create a WiFi access point (AP).

To configure WiFi access so that it can connect to your home network, you disconnect your PC, tablet, or phone from your usual WiFi network and connect to the Raspberry Pi access point instead.

Then you configure the Raspberry Pi for your home network and reboot the hotspot. Lastly, you reconnect your PC, tablet, or phone to your home network.

1. Boot up the hotspot and wait for two or three minutes until the AP has been created.
2. On a Windows PC, click the WiFi icon on the right side of the toolbar. On a phone or tablet, the WiFi settings are in the settings menu. Look for a new WiFi network called **Pi-Star-Setup and** connect to that. I did not have to enter a password, but if you do, it will be, **raspberry**.
3. You should be taken directly to the Pi-Star dashboard. If that does not happen, open your web browser and enter pi-star.local into the URL area. After a few seconds, the Configuration page should appear. Or click on **Configuration** top right.

TIP: Your web browser may want to treat pi-star.local as a search enquiry. If that happens enter http://pi-star.local or http://pi-star instead.

Thanks to W1MSG for his video at https://www.youtube.com/watch?v=Z5svLP8nEyw

Skip ahead to the WiFi configuration instructions on page 134.

Option 2: Adding the WiFi settings using a manual setup

Pop back to the Pi-Star.uk website and select **Pi-Star Tools > WiFi Builder**. Enter your country code using the WiFi **Country Code** dropdown list. In the **SSID** box enter the name of your WiFi Network. In the **PSK** box enter your WiFi password, the same as if you were connecting a new phone or laptop to your network at home. Then click **Submit**.

TIP: If you require a config that will connect to any available open network, leave the SSID and PSK lines empty, the generated config will allow your Pi to connect to any available open network. Then click on **Submit**.

The website will download a small file called wpa_supplicant.conf.

Open Windows Explorer and look at the boot drive on the SD card. There will be a second drive as well, but you can ignore that. Copy or move the wpa_supplicant.conf file into the SD card boot drive partition.

TIP: The SD Card boot drive should open when you plug the SD card (in its USB adapter) back into the PC. Close or cancel any Windows error popups and do not accept the instruction to format the drive partitions.

Done! Take the SD card out and insert it into the SD card slot on the Raspberry Pi.

The card will only go in one way, copper terminations in first, with the card facing "up" towards the printed circuit board. It will slide in easily. There is no latch.

1. When the card has been inserted into the Raspberry Pi and the Pi-Star system boots up, it will add the config file for the WiFi and then reboot. There may be no indication on the hotspot that this has happened, and it may take a couple of minutes to sort itself out.

2. Open the web browser on your PC and enter pi-star.local into the URL area. After ten seconds, the Configuration page should appear. Or click on **Configuration** top right.

TIP: Your web browser may want to treat pi-star.local as a search enquiry. If that happens enter http://pi-star.local or http://pi-star instead.

Skip ahead to the Pi-Star Configuration chapter.

Option 3: Adding the WiFi settings using the wired Ethernet method

If you have a Pi3 or 3B you can temporarily connect an Ethernet cable between your LAN router and the hotspot. If you have a situation where the only LAN port is where the fibre or ADSL modem is, and it is remote from the shack, that is not a problem. You can configure the hotspot with the hotspot and the PC separated provided they are on the same LAN. After the WiFi has been configured you can unplug the hotspot and move it back into the shack.

Open the web browser on your PC and enter pi-star.local into the URL area. After a few seconds, the Configuration page should appear. Or click on **Configuration** top right. If you cannot connect to the hotspot, you will have to find the IP address of the hotspot and enter that. You can find the hotspot listed as 'Pi-Star' by accessing your main router, or by using the NetAnalyzer (iPhone) or FING app.

TIP: Your web browser may want to treat pi-star.local as a search enquiry. If that happens enter http://pi-star.local or http://pi-star instead.

Continue to the WiFi configuration instructions.

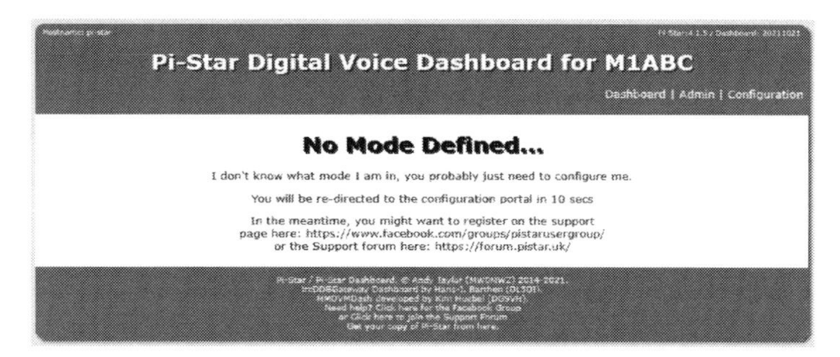

Figure 36: Initial Pi-Star screen

WIFI CONFIGURATION INSTRUCTIONS

If you used option 1 or option 3, you have not finished yet. If you used option 2 the hotspot should already be configured for WiFi, and you can skip this section and proceed to the Pi-Star Configuration chapter.

When you power up the hotspot you should see the 'No Mode Defined' Pi-Star splash screen and after 10 seconds the Configuration page should open. If it did not open, click Configuration at the top left of the Pi-Star screen. I will cover the general Pi-Star configuration in the next chapter. This section is just for completing the WiFi setup.

Scroll down the Config screen until you reach the 'Wireless Configuration' section. It should look like this. Note that it says that the WiFi interface is down. If it says the WiFi interface is up, the WiFi is already configured and working.

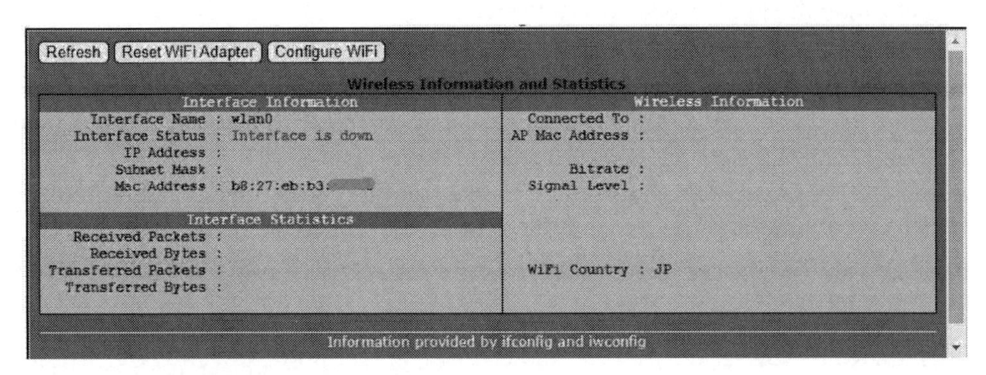

Figure 38: Pi-Star WiFi status

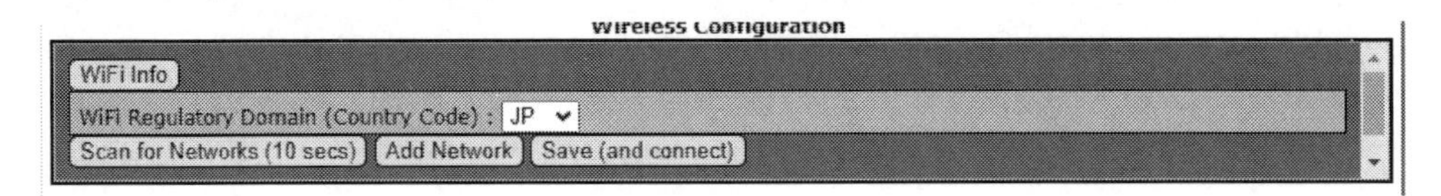

Figure 38: Pi-star WiFi Config.

Just below the status box, there should be a configuration box, like this one. Select your country using the **WiFi Regulatory Domain (Country Code)** dropdown list. I selected NZ. Then press the **Scan for networks (10 seconds)** button. It should provide you with a list of WiFi sources that the hotspot can see. Your home LAN should be near the top because it should have the strongest signal. Select your WiFi LAN. Alternatively, you can set this up manually by pressing **Add Network**.

This should populate the Wireless Config with your **Country Code** and **SSID** (WiFi network name). Enter your **WiFi password** into the PSK box and press **Save (and connect)**.

Wireless Configuration

WiFi Info

WiFi Regulatory Domain (Country Code) : NZ

Network 0 Delete

SSID : SPARK-DL

PSK :

Scan for Networks (10 secs) Add Network Save (and connect)

Networks found :

Connect	SSID	Channel	Signal	Security
Select	SPARK-DLZADR	2.4GHz Ch8	-69 dBm	WPA/WPA2-PSK (TKIP/AES) with WPS

Figure 39: Choosing the WiFi network and entering your WiFi password

Above the 'Wireless Communication' box, there is a button marked **Apply Changes**. Click that. After about 30 seconds the Config screen will go back to the splash screen and then after another 30 seconds or so, it will return to the config screen. You should be rewarded with the green **Interface is up** message.

Take note of the IP Address in case you need it later.

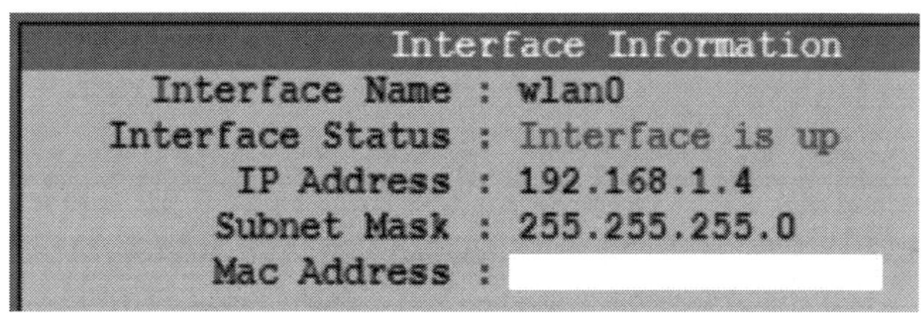

Figure 40: The WiFi is working :-)

IMPORTANT NOTE

Now you need to restart the hotspot to transfer from the AP mode (option 1) or the Ethernet mode (option 3). At the top of the page, select **Power > Reboot > OK**. Wait 90 seconds for the reboot to finish, and the dashboard should reset.

If you used option 1, now is the time to reconnect your phone, PC, or tablet to your normal network. If you used the Ethernet cable and you want to swap over to WiFi, unplug the Ethernet cable before you reboot the hotspot.

The hotspot will jump to a different IP address when it restarts. It should be the one you recorded. That means that your Pi-Star web page may not work anymore. Retry the pi-star.local or http://pi-star.local page several times and it should eventually restart. If that does not work, try entering the IP address as a URL. See the next section.

Pi-Star configuration

OK, the hotspot is running, and you have configured the WiFi access, which is "the tricky part." Now Pi-Star has to be configured.

This chapter covers the initial Pi-Star configuration in detail. If you already have your Pi-Star configured and working fine based on the information in the Pi-Star dashboard chapter, you can skip this chapter.

Pi-Star runs on the Raspberry Pi, but you configure it using a webpage interface on your PC, tablet, or phone. This is normally only accessible from within your home network. You can choose to make it public, but that would only be for a public repeater or high-powered hotspot.

If you are not already there, with the hotspot booted up and running and Pi-Star on your web browser, plug in the hotspot, start your PC web browser, and type

http://pi-star/ or pi-star or

http://pi-star.local/ or pi-star.local or

http://pi-star/admin/configure.php or

the IP address, usually something like 192.168.1.4

At least one of these options should bring up the Pi-Star dashboard.

PI-STAR ERROR MESSAGE

You may get an error message like this one. It just means that the modem selected in Pi-Star does not match your hotspot hardware, which is not surprising since you have not set that yet. Just click OK and carry on.

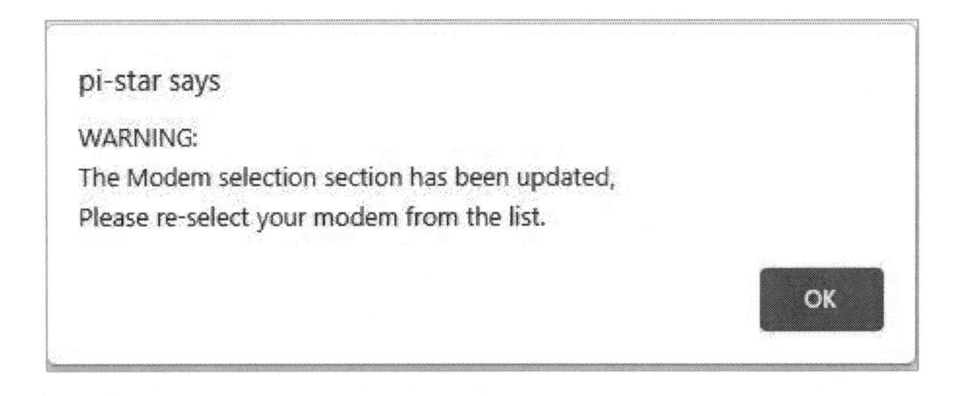

GATEWAY, HOST, AND GENERAL CONFIGURATION

If you are on the dashboard page, click **Configuration** to get started. The username is **pi-star**, and the password is **raspberry**.

Wow, this looks complicated! Luckily you only need to do a few things. D-Star configuration is much easier than DMR.

Gateway Hardware Information

Hostname	Kernel	Platform	CPU Load	CPU Temp
pi-star	5.10.63-v7+	Raspberry Pi 3 Model B Rev 1.2	0.44 / 0.74 / 0.85	47.2°C / 117°F

The top section is 'Gateway Hardware Information.' It contains the hostname, software kernel (revision number), the platform (model of Raspberry Pi), CPU loading, and the CPU temperature. Green is good. Orange is OK. If it gets into the red, you need a heatsink or a fan on your Raspberry Pi CPU.

Control Software

Setting	Value
Controller Software:	DStarRepeater / MMDVMHost (DV-Mega Minimum Firmware 3.07 Required)
Controller Mode:	Simplex Node / Duplex Repeater (or Half-Duplex on Hotspots)

Apply Changes

The next section is 'Control Software.' It should already have **MMDVMHost** selected because we are not configuring a D-Star repeater. If you have a simplex hotspot (one antenna) select **Simplex Node**. If you have a duplex hotspot (two antennas) select **Duplex Repeater**. This is **an important setting.**

MMDVMHost Configuration

Setting	Value		
DMR Mode:		RF Hangtime: 20	Net Hangtime: 30
D-Star Mode:		RF Hangtime: 20	Net Hangtime: 20
YSF Mode:		RF Hangtime: 20	Net Hangtime: 20
P25 Mode:		RF Hangtime: 20	Net Hangtime: 20

The third section, 'MMDVM Host Configuration,' is where we set up different digital voice modes and transcoding between voice modes. We are only interested in D-Star at present, so click the switch icon to turn **D-Star Mode** on. The switch should change to an orange colour. We will leave the **RF Hangtime** and **Net Hangtime** set to the default 20 seconds. You can change them later if you want to.

You can select other modes such as YSF or DMR if you have radios for those modes. But for now, it will be less confusing if you just enable D-Star. It can be confusing when calls come in and you could miss part of a conversation on one mode if the hotspot switches to a different mode. The hang time settings create time (20ms) for someone to come back to your call before a call on another mode can take over. If you are using multiple modes and this becomes a problem, you can increase the hangtime to alleviate the issue.

MMDVDM Display Type

The last line in this section selects the display type if your hotspot has a display. The MMDVM Display Type is set according to what display (if any) you have on your hotspot.

If your hotspot has no display, set the MMDVDM Display type to **None**.

For an OLED (small display) select OLED type 3 for the very small 0.96′ screen or **OLED type 6** for the more common 1.3″ screen. Either option works on my hotspot.

If you have a Nextion screen, select **Nextion**.

There are a couple of other options, including one for TFT displays.

Port

For OLED displays, select **/dev/tty/AMA0** because the display is being driven directly by the modem board.

For Nextion displays you usually select **modem**. However, if the Nextion display is connected to a TTL to USB adapter plugged into the Rpi, you should select **/dev/tty/USB0**.

Nextion Layout

This selects from four display layout options. Note that the background image is stored on the display itself. This dropdown only changes the information that is sent from Pi-Star to the display. It does not change the basic layout. Setting the Nextion screen resulted in my display starting to show the IP address.

If you have or can download the HMI file, you can edit the Nextion layouts on your PC using the free Nextion Editor and load it directly back into the Nextion display. You may have seen a power adapter in the Nextion box. See the information at. https://on7lds.net/42/nextion-displays and https://nextion.tech/nextion-editor/.

Click the **Apply Changes** button to load your changes to the hotspot.

Apply changes

You will see **Apply Changes** at the bottom of each section. It should be used every time you make changes in any of the configuration sections. New options often appear on the configuration forms after you click **Apply Changes**. You should click the button and wait for the hotspot to reboot, then check what changed before proceeding to the next section.

General Configuration

The 'General Configuration' section includes some very **important settings**.

The **Hostname** is **pi-star**. I do not see any good reason to change this.

The **Node Callsign** is usually your **callsign**. If it was a public access repeater, it would be the repeater callsign.

Frequencies

If you selected 'Simplex Node' there will only be one frequency box. If you entered 'Duplex Repeater' and remembered to Apply Changes, there will be a receive frequency and a transmit frequency.

Remember that the duplex hotspot frequencies are the reverse of the frequencies that you program into your radio. The radio receives the hotspot's transmitter (output) frequency, and it transmits on the hotspot's receive (input) frequency. I know that you already know this, but it is easy to get it wrong 😉.

You can use any frequencies that are in the band(s) supported by both the hotspot and your handheld radio. Almost all MMDVM hotspots operate in the 70cm amateur band. Although the hotspot transmits very low power and is unlikely the interfere with others, you must remember that your handheld or mobile radio may be transmitting a much higher power. Check your local 70cm band plan. There will be band segments for repeaters and for digital simplex. For a simplex hotspot, I would choose one of the designated digital simplex frequencies or a repeater output frequency that is not used in your area. I believe that it is a good idea to use the standard band segment and ± 5 MHz repeater offset for a duplex hotspot on the 70cm band, or ± 600 kHz offset on the 2m band. I selected a repeater pair that is not in use for any other repeaters in New Zealand. Just choose repeater frequencies that are not being used in your region. If somebody complains it is a trivial matter to change the hotspot to a different frequency, although it will mean changing the hotspot channel on your radio.

General Configuration

Setting	Value	
Hostname:	pi-star	Do not add suffixes such as .local
Node Callsign:	ZL3DW	
Radio Frequency RX:	433.125.000	MHz
Radio Frequency TX:	438.125.000	MHz
Latitude:	-43.497	degrees (positive value for North, negative for South)
Longitude:	172.605	degrees (positive value for East, negative for West)
Town:	Christchurch, RE66hm	
Country:	New Zealand	
URL:	https://www.qrz.com/db/ZL3DW	◉ Auto ○ Manual

Figure 41: Pi-Star identity and frequencies

Location

Add your **latitude and longitude,** using the 'degrees and decimal' 172.1234 notation rather than 'degrees minutes and seconds.' If you don't know your latitude and longitude, find your street on Google maps click on the map, then right-click, and your location will be displayed. The latitude is a positive number in the northern hemisphere and a negative number in the southern hemisphere. The longitude is measured in degrees East (+ve) or degrees West (-ve) of the zero longitude line.

Enter your closest **town** or your locality and your Maidenhead grid. For example, Christchurch, RE66hm. Enter your **Country**.

For **URL** you can enter your personal website, but most folks select **Auto** which automatically uses your QRZ.com listing. Note that selecting Auto won't change the URL text box until you go through the **Apply Changes** process.

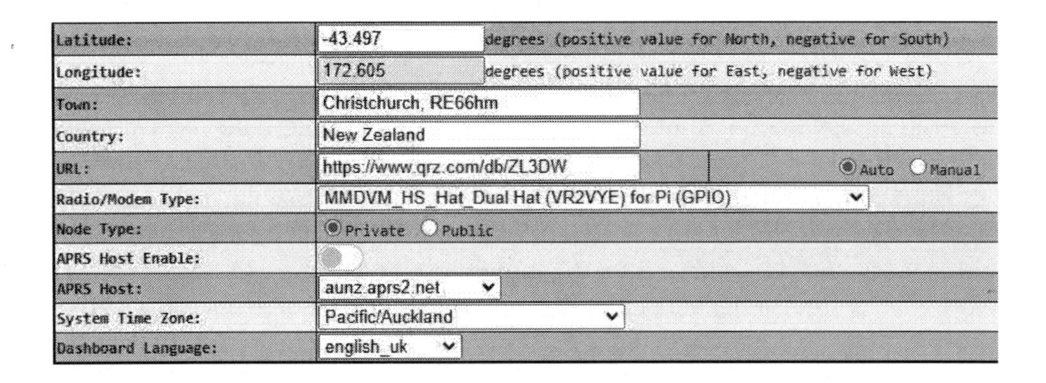

Radio modem type

The radio modem type has to suit the hotspot design. You might have to experiment to find the correct one if the manufacturer didn't supply the modem type information. It will be an RPi one, not a Nano or a USB stick. Mine is a 'dual hat' (duplex) BI7JT hotspot from China. The firmware was written by VR2VYE.

The STM32-DVM / MMDVM_HS – Raspberry Pi Hat (GPIO) modem seems to work with most generic Chinese simplex hotspots. If you bought a ZUMspot try one of those options, if it is a DV-Mega try one of those options.

Node type

Set the node type to **Private**. It would only be set to public if your repeater or hotspot was also available publicly. If you do set it to 'public,' change the Remote Access Password at the bottom of the screen.

APRS Host

If you turn **APRS Host Enable** on, set the closest APRS host, or rotate.aprs2.net.

This setting lets the hotspot announce its position on the aprs.fi website when it boots up. It also passes D-PRS position information to aprs.fi when you transmit. I leave it turned off since my hotpot and radio are not travelling anywhere.

System time zone and language

Set the time zone and language to your time zone and preferred language option.

Click **Apply** Changes when you have finished with the General Configuration settings.

D-STAR CONFIGURATION

The 'D-Star Configuration' section sets the D-Star network settings. This is much easier for D-Star than it is for DMR. The section will be displayed once you have selected the D-Star mode on the MMDVM Host Configuration and clicked **Apply Changes**.

RPT1 Callsign is the callsign of the hotspot (or repeater) node. The callsign was set in the General Configuration section, but you can add a module code here. Normally the module code relates to the frequency of the hotspot or repeater. A = 23cm, B = 70cm, or C = 2m. But if you already have a hotspot using that module code, you can select any other code except G. You cannot have two hotspots using D-Star with the same module code.

RPT2 Callsign is the gateway callsign of the hotspot (or repeater) node. It is always set to the node callsign with a G module code.

The **Remote Password** is the **raspberry** password required to access the Admin and Configuration pages. Don't change it unless you are making the hotspot public access. If you are making hotspot public access, you should definitely change the remote password to stop Internet users from changing your hotspot or repeater configuration.

The **Default Reflector** is the reflector that the hotspot will connect to when it is booted or restarted with the Apply Changes button. Set it to the reflector that you use the most. Select **Startup**. If you select Manual, the hotspot will not connect to a reflector on startup. It will stay 'Unlinked' and you will have to establish a link manually.

The **ircDDBGateway** language defaults to English but there are many other settings. **Time announcements** enables hourly time announcements via the radio. It might be handy if you have a sked, or like me, you need reminding when to go to bed.

Callsign routing establishes a connection via the ircDDBGateway code in Pi-Star. You don't need this for normal reflector or gateway linking. I don't use it. The Callsign Routing option is supposed to enable calling an individual callsign or routing to 'Smart Routing Groups,' usually referred to as' QuadNet smart groups' or 'STARnet groups.'

If you have 'Callsign Routing' enabled there will be a line showing the QuadNet link, ircv4.openquad.net on the D-Star Network section at the bottom left of the Pi-Star dashboard. If it says rr.openquad.net the Pi-Star needs an **Update**, or you can change it in the Expert settings.

Use DPlus for XRF, makes connections to XRF reflectors using the DPlus protocol, the same as REF reflectors, rather than the newer DExtra protocol. All XRF reflectors accept either protocol. Some home networks don't play well with DExtra because it requires port forwarding of port 30001 in your Internet router. Enable this function if you find that you are unable to link to XRF reflectors. If you change this, you have to do a Pi-Star update as well.

MOBILE GPS CONFIGURATION

This setting has nothing at all to do with the GPS receiver on the radio. It is used when you connect a GPS receiver dongle to the Raspberry Pi that is hosting the MMDVM hotspot. You can turn on the serial port or USB port if a USB dongle is plugged in and the baud rate. Normally 4800 or 9600 for a GPS receiver but may be faster for a dongle.

PI-STAR FIREWALL CONFIGURATION

The firewall configuration section is for experts. You can leave everything set to the default options. All the access settings should be **Private**, **Auto AP** should be on, and **uPNP** should be on.

If you need access over the Internet, you can change any or all of the top three settings to Public. These settings are used for accessing the dashboard remotely, from outside your network. To quote Andy Taylor in the Pi-Star Users Support Group: "These settings tell the uPNP daemon to request port forwards from your router."

- Dashboard Access: requests TCP/80
- ircDDBRemote Access: requests UDP/10022
- SSH Access: requests TCP/22

Auto AP lets the Raspberry Pi act as a WiFi access point if it is unable to find your WiFi network within 120 seconds. You can link your phone to the node and configure the WiFi access on the Raspberry Pi.

uPNP lets the Raspberry Pi manage its own firewall settings. If you turn it off, you can configure the firewall settings in the Expert tab.

Setting Dashboard access to public would allow Internet users to observe the Pi-Star dashboard. This is OK providing you changed the 'Remote Access Password' to stop them from changing the configuration settings. Making the ircDDBGateway public would allow others to manage routing. Changing SSH access enables remote users to access the Raspberry Pi via Secure Shell. This could be required if someone is offering remote support to fix a Raspberry Pi configuration issue. Note that SSH access within your LAN is always available. You can access the Raspberry Pi via the puTTY SSH (secure shell) program without changing this setting.

Wireless Information and Statistics

Wireless Information and Statistics shows the WiFi connection. It should state 'Interface is Up.' You can configure the WiFi if it is not working.

AUTO AP SSID

This setting changes the default Pi-Star identification so that a password and login name are required to access the hotspot in AP mode. If you need the AP mode, you are probably in enough trouble without adding this extra layer of complexity. Leave both entry boxes empty.

REMOTE ACCESS PASSWORD

The remote access password should only be set if you are making the hotspot public. You would normally only do that if you were using Pi-Star on a repeater, and you wanted to make the dashboard available for repeater users. If you are the only person that needs to see the hotspot dashboard, keep it set to private, and do not enter a remote access password.

POWER

Clicking the **power** menu bar item at the top of the page provides this attractive image. Click the green **reboot** icon to restart the hotspot or the red **shutdown** icon to safely shut down the hotspot. The file system on the Raspberry Pi SD card can become corrupted if you just pull the power plug out. It is always best to shut down the Pi using this power button, or a button on the hotspot display if it has one (mine does not). I admit that I usually just pull the power plug. But that is a risk that I take, and I can restore the software if I have to.

PI-STAR BACKUP

When you have your settings right, or nearly right, make a backup of the Pi-Star configuration. It will be saved to your Windows downloads area as a Zip file. Click **Configuration > Backup/Restore > Big down arrow.**

You can restore a previous arrangement by selecting the file with **Choose File** and clicking the **Big up arrow.** This could be useful if you make configuration changes that didn't work well, or if you want to save multiple arrangements. Perhaps separate DMR and D-Star configurations. You can also copy a zip file to load a Pi-Star configuration to or from another hotspot.

UPDATING PI-STAR SOFTWARE AND MMDVM FIRMWARE

WARNING: There is no need to update the Pi-Star software if you just downloaded it from the Pi-Star.uk website. It will already be the latest version. There will probably be no need to update the MMDVM hotspot firmware either unless it is a very old modem. Don't update either unless you have a good reason to do it.

Pi-Star software update

You can update Pi-Star from the Pi-Star dashboard. The version you are running is displayed at the top right of every dashboard page. You probably never noticed it. My dashboard has 'PI-STAR: 4.1.6 / Dashboard: 20220904. That means that the Pi-Star version is 4.1.6 and the Dashboard version was released on the 4th of September 2022.

To update the software, click **Configuration > Update.** A load of green text will show the Linux update happening on the Raspberry Pi. When it has finished, re-boot the hotspot and if there was an update available, the dashboard page will show the new version number.

WARNING: if you click UPDATE, there is no backing out. The update will begin immediately. It takes several minutes to run. Do not interrupt the update while it is running. The hotspot will reboot at the end of the update. "Patience Grasshopper."

SSH access to Raspberry Pi

You can access the Raspberry Pi that is running your hotspot with SSH (Secure Shell). Most people use PuTTY for SSH access, but any SSH client will do.

I can't think of any reason to access the Raspberry Pi, but you can do a software update that way if you want to.

Enter the Pi-Star's **IP address** or **pi-star.local.** into the host name box. Save it see the note below. Then click **Open.** I found that using the host name crashed PuTTY, but the page would come up after a delay of a minute or more. It is probably something to do with my PC setup. Using the IP address worked perfectly.

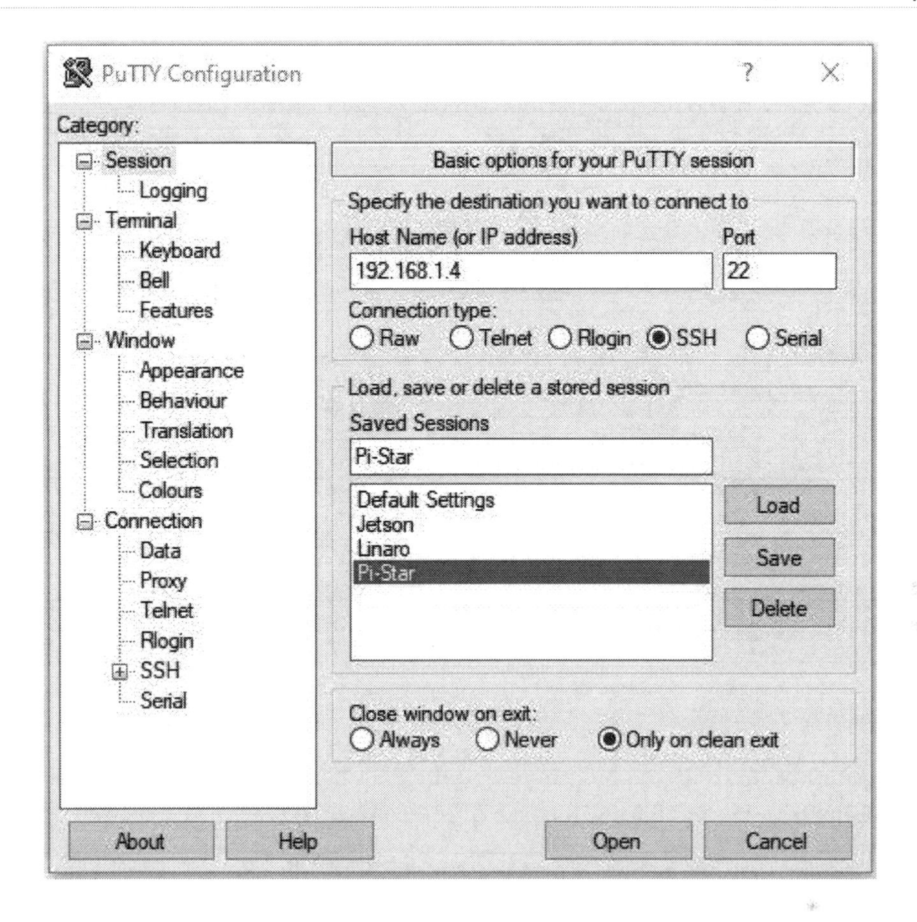

It is a good idea to enter the hostname or IP address and then save it by entering a name in 'Saved Sessions' and clicking the Save button. If you don't, this rather ominous warning message will pop up when you click Open. Just ignore it and click **Yes**.

You will be presented with a PuTTY terminal window.
login as **pi-star**
pi-star@pi-star.local.'s password is **raspberry**

Figure 42: SSH access to the Raspberry Pi

This screen indicates that you have access to the Raspberry Pi. You can use standard Linux commands, **ls -a** to list a directory, **df -h** to check the SD card space, etc.

Note the instructions on the boot screen that say 'Pi-Star's disk is read-only by default. You can change it to read-write access at your own risk, using **rpi-rw**. You do not need to do this to run the update or upgrade commands. Also, note that all of the Pi-Star tools start with a **'pistar-'**, prefix. There is no hyphen between pi and star.

TIP: You can also access the Raspberry Pi by plugging in a USB keyboard and an HDMI monitor, and then rebooting the hotspot.

Updating the Pi-Star software from the command line

There is no need to update the Pi-Star software from the command line because you can do it from the Pi-Star dashboard. But if you feel the need, you can, using **sudo pistar-update**. I believe that it runs the same command as doing it from the dashboard, but it also updates the Raspberry Pi OS and other software on the Raspberry Pi. It can take a long time to complete. The update will stop for up to a minute on some lines. Just be patient, go and make a coffee, and let it run until you get the Linux prompt back.

You can also run **sudo pistar-upgrade** which will look for an upgraded version of Pi-Star. Well, that was interesting! When I ran the upgrade command, it said I was already running the latest version (as expected) but it now the dashboard reports that I am running version 4.1.6.

Updating the MMDVM firmware from the command line

The MMDVM firmware is indicated in the Radio Info box on your Pi-Star dashboard. It is updated using a command like **sudo pistar-mmdvmhshatflash hs_hat** or if you own a ZumSpot **sudo pistar-zumspotflash rpi**. You **MUST** know the correct firmware for your hotspot. It should be on the manufacturer's website. **If in any doubt don't do it**.

Attempting to load firmware that is not correct for the hotspot that you own can turn your hotspot into a wafer of unusable silicon. Often known as a 'brick' since it is as useless as a brick. This situation could be salvageable if you try again with the correct firmware, or you might have to buy a new hotspot.

LIVE LOGS

The Live logs menu option opens a screen that logs each incoming and outgoing call until you exit the screen. I suspect that it is Linux data from the Raspberry Pi since the text is green. The log of incoming calls includes text like 'received network header,' or 'received network end of transmission.' RF signals into the hotspot create 'RF header,' or 'RF end of transmission' messages.

I expect the function is for fault finding, but you could use it to find out if a certain station called or activated the reflector while you were away from your desk. You can save the file as a comma-delimited text file by clicking the word **'here'** beside 'Download the log' at the bottom of the screen. Once the file has been saved, you can open it file with Excel, Word, or a text editor such as Notepad or Wordpad.

Technical information

D-Star is an open standard specifically designed for amateur radio. It has been fully integrated by Icom but very few other manufacturers have adopted it. There used to be some Kenwood models including the excellent TH-D74 handheld radio. For the normal D-Star DV (digital voice) mode, the speech audio is encoded as a 3600 bps (bits per second) data stream using AMBE+2 encoding. The 3600 bps stream includes 2400 bps of voice data and 1200 bps of FEC (forward error correction). There is an additional 1200 bps for data transmission, resulting in a total data rate of 4800 bps. The data bits are used for synchronisation, and they carry messages like your callsign, location, radio type, the station, gateway or reflector you are connecting to, and the repeater you are using. The D-Star digital voice (DV) mode is commonly used on the 2m, 70cm, and 23cm bands.

There is also a 'high speed' Digital Data (DD) mode which can be used to send data at 128 kbits. This is limited to the 23cm (1.2 GHz) band and above because of the 150 kHz channel bandwidth required for the transmission.

VOICE CODING

When you transmit, your voice is converted into a digital data stream of binary bits which is coded with the AMBE+2 vocoder in the radio and then applied to the GMSK modulator in the transmitter. The coding conversion is done with a propriety chip running the DVSI (Digital Voice Systems Incorporated) AMBE+2 Vocoder. A Vocoder is a CODEC (coder-decoder) designed for voice signal digital coding, compression, multiplexing, and encryption. Initially, D-Star used the older AMBE vocoder, but now, the three most popular digital voice modes all use the AMBE+2 Vocoder.

In the receiver, the GMSK signal is demodulated back to a data stream and the AMBE+2 Vocoder converts it back into an audio signal, ready for the audio amplifier and speaker, Bluetooth headset, or headphones.

MODULATION

D-Star uses GMSK (Gaussian Minimum Shift Keying) which is a variation of frequency shift keying (FSK). Frequency shift keying alters the frequency of the transmission to represent changes in the binary data between the digital one level and the digital zero level.

Minimum Shift Keying (MSK) limits the amount of the frequency shift. It has a modulation index of 0.5 indicating that the difference between the two frequencies is equal to one-half of the bit rate. Unlike PSK, the transitions are managed to that there is a smooth change between frequencies. This reduces the bandwidth of the signal compared to PSK at the same data rate. In GMSK, the modulating data signal is passed through a 'Gaussian' low-pass filter before the modulation process.

The filter rounds the edges of the data pulses, further improving the smoothness of the transitions between the frequency representing the digital one level and the frequency representing the digital zero level. This technique narrows the transmitted bandwidth a little more, reducing the bandwidth of the D-Star transmission to 6 kHz so that it fits within a 6.25 kHz channel spacing. That is half the 12.5 kHz channel spacing used for analog FM. D-Star is the only amateur radio digital voice mode that fits into a 6.25 kHz channel, although DMR fits two voice channels into a 12.5 kHz channel.

GMSK was used in the early 2G GSM cellular radio systems. A Gaussian filter has a filter characteristic that approximates a Gaussian distribution. The shape is a 'bell' curve.' The mathematics is a bit complicated and not necessary here. Gaussian distribution is named after the German mathematician Carl Gauss who first described it. The chief benefit of a Gaussian filter is that it does not exhibit ringing.

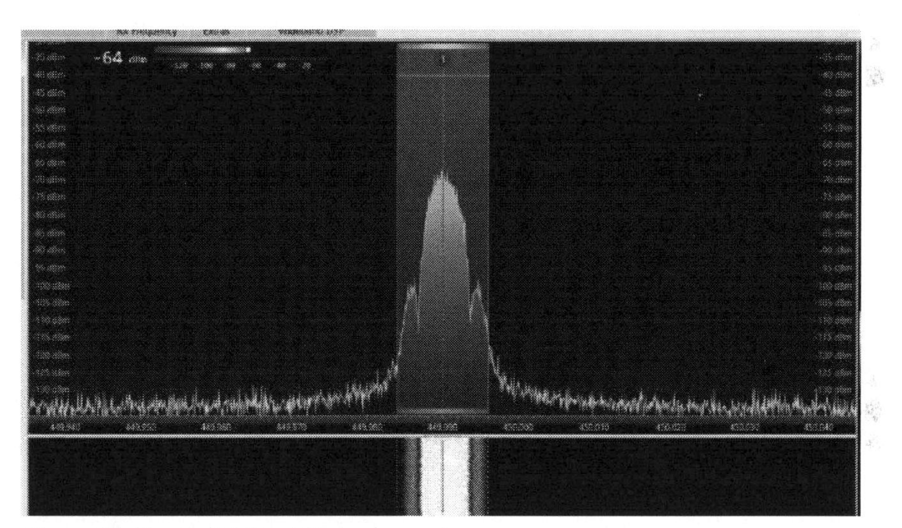

Figure 43: Typical D-Star DV mode spectrum

Image from: https://robrobinette.com/Ham_VHF-UHF_Digital_Modes.htm

The developers decided to use GMSK rather than QPSK because although GMSK is not quite as sensitive as QPSK, it uses less bandwidth for a similar data rate, and it is less affected by frequency errors between the transmitter and receiver. Even with a 0.5 ppm TCXO, there can be quite big frequency errors on a 1.2 GHz signal. The 4800 bps data rate was chosen for the DV mode because it is the maximum speed that can be achieved within a 6 kHz bandwidth. The 128 kbit data rate is acceptable for the DD mode on the 1.2 GHz band because there is no requirement for 6.25 kHz or 12.5 kHz channel spacing. The 23cm (1.2 GHz) band is 60 MHz wide, compared to 10 MHz on the 70cm band and 4 MHz on the 2m band, in ZL and the USA, or 2 MHz in the UK and Europe.

THE DIFFERENCE BETWEEN FSK AND PSK

With frequency shift keying (FSK) the binary data signal is represented by changes in the RF frequency. In phase-shift keying (PSK) the frequency remains the same and the binary data is represented by changes in the phase of the signal. Each phase or frequency state is known as a symbol. The symbol rate for 2-FSK (FSK) or 2-PSK (BPSK) is the same as the bit rate because the two possible frequency or phase states, each represent one binary bit. Note that it is not the current frequency or phase that determines whether the symbol is a binary one or a zero. It is the change of state. There are scrambling and data coding techniques aimed at eliminating long runs of ones or zeros.

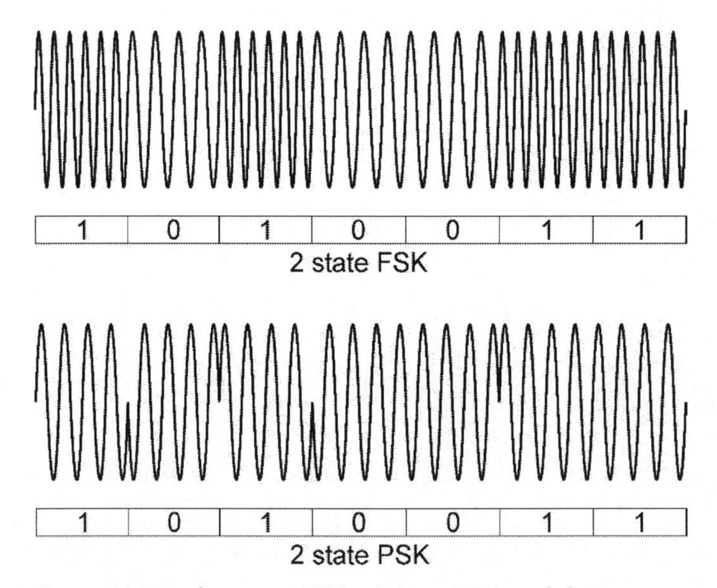

Figure 44: Two frequency FSK vs Binary PSK modulation

Many digital systems use differential coding where a change of state means that the current binary bit is different to the previous bit. No change of state indicates that the current binary bit is the same as the previous bit.

4FSK (four-frequency shift keying) uses four frequencies rather than two frequencies. 4PSK (four-phase shift keying), also known as QPSK (quadrature phase-shift keying) uses four phase changes, 0°, 90°, 180°, and 270°, rather than the 0° and 180° phase shifts used in 2PSK also known as BPSK (binary phase-shift keying).

With 4FSK and QPSK, each phase or frequency state carries 2 bits of the digital signal, so the bit rate is twice the symbol rate. Using this higher order of modulation allows you to transmit twice as much data in the same amount of time. The data is usually manipulated so that all four states are transmitted regularly. This is particularly important for QPSK because it affects the shape of the transmitted spectrum, and clock recovery at the receiving station.

THE DV (DIGITAL VOICE) DATA FORMAT

The overall data rate for the DV (digital voice) mode is 4800 bits per second.

- 2400 bps for the voice channel
- 1200 bps for forward error correction (FEC)
- 1200 bps for sync and signalling data

D-Star DV voice packet 4800 bps

Radio packet header				Voice and data packet																					
Sync	Flags	Identification	Chk	V	D	V	D	V	D	V	D	V	D	V	D	V	D	V	D	V	D	V	D	V	D
10 -1bit	3	36	2	72	24	72	24	72	24	72	24	72	24	72	24	72	24	72	24	72	24	72	24	72	24

The packets consist of a header which includes a packet header checksum, followed by voice and data frames. Finally, there is a data checksum to indicate whether the receiving station has received the voice and data frames accurately.

After the packet header, described below, there are 11 pairs of voice and data frames. The 'V' (voice) frames are 72 bytes long, and the 'D' (data) frames are 24 bytes long. The last data package of a transmission, not a packet, contains synchronisation data.

THE DD (DIGITAL DATA) FORMAT

The DD mode takes Ethernet data and packages it for transmission over the radio. This has the advantage that Internet or LAN data can be transmitted without modification. That makes the radio path look transparent to the computers at either end. It is suitable for transmitting files, pictures, streaming audio, emails, or even web pages. The data rate for the DD (digital data) mode is 128 kbits per second. It requires a bandwidth of 130 kHz which fits within a 150 kHz channel spacing.

D-Star DD data packet 128 kbits

Radio packet header				Ethernet data packet	FCS
Sync	Flags	Identification	Chk	66 to 152 bytes	CRC

The packets consist of a header, followed by an Ethernet packet which includes a CRC-32 (cyclic redundancy check) checksum of the Ethernet data. The Ethernet data can be anywhere from 66 to 1520 bytes long including the checksum data.

Each Ethernet data packet includes 2 bytes to indicate the frame length, destination and source MAC address (6 bytes each), Ethernet type (2 bytes), 46 to 1500 bytes of data, and the 4-byte CRC error check.

THE PACKET HEADER

The packet header is the same for both modes. It starts with bit and frame sync data. The 'bit sync' is always the same. It is the standard 64 bit pattern used by all GMSK systems. The frame sync is unique to D-Star. It is always the same 1110 1100 1010 000, 15-bit pattern.

D-Star packet header

Sync		Control			Identification					Chk
Bit sync	Frame sync	Flag 1	Flag 2	Flag 3	Destination Repeater Call Sign	Departure Repeater Call Sign	Companion Call Sign	Own (your) Call Sign 1	Own (your) Call Sign 2	P-FCS Checksum
64 bits	15 bits	1	1 bytes	1	8 bytes	8 bytes	8 bytes	8 bytes	4 bytes	2 bytes

The frame sync is followed by three 1-byte flags. Flag 1 indicates whether the data being sent is control data or user data and whether the communications are via a repeater or simplex, plus some priority status bits. Flag 2 is unused. Flag 3 is reserved. It will be used if the D-Star protocol is updated, to indicate which protocol is being used. The next 36 bytes carry the 'Identification data.' They are:

- Destination repeater callsign (up to 8 ASCII text characters). It is the repeater the far station is using. Usually, a six-digit callsign a space and a service identifier or 'module' (A, B, C etc.) For a simplex call, it changes to DIRECT.
- Departure repeater callsign (up to 8 ASCII text characters). It is the repeater you are using. Usually, a six-digit callsign a space and a service identifier or 'module' (A, B, C etc.) For a simplex call, it changes to DIRECT.
- Companion Callsign (up to 8 ASCII text characters). The station you are calling. Usually, a six-digit callsign a space and a service identifier or 'module' (A, B, C etc.). If you are not linking to a reflector or sending I, U, or E, the 'companion callsign' will be 'CQ CQ CQ.'
- Your callsign 1 (up to 8 ASCII text characters). Usually, a six-digit callsign a space and a service identifier, module A=2m, B=70cm etc.
- Your callsign 2 (up to 4 ASCII text characters) contains the four letters of suffix data that follows the /. For example, '9700' if you are using an IC-9700.

Finally, there is a 2-byte checksum calculated from the flag and ID bytes.

COMPARISON OF DIGITAL VOICE MODES

All the digital voice systems in common use on the UHF and VHF amateur radio bands use digital forms of FM modulation. This is because it has been traditional to use FM at those frequencies and because commercial operators are interested in migrating customers from FM repeaters onto digital repeaters. It is good because it means that all digital voice radios support FM operation and can access FM repeaters.

The 4FSK modulation used for DMR and the C4FM modulation used for YSF produces a four-state constellation similar to QPSK. The bit error rate (BER) performance of the two modes is nearly identical (within about 1 dB). The GMSK modulation used for D-Star is slightly better. DMR requires a received signal two or three dB stronger to achieve the same BER as D-Star.

All the modes in this comparison use extra transmitted bits to perform FEC (forward error correction) which improves the voice quality by repairing errors in the received data.

RF factors	DMR	D-Star	Fusion	P25 phase1
Vocoder	AMBE+2	AMBE+2	AMBE+2	IMBE
FEC	Yes	Yes	Yes	Yes
Modulation	4FSK	GMSK*1	C4FM*2	C4FM*2 or CQPSK
Emission	7K60FXE	6K00F7W	9K36F7W	8K10F1E
Transmission rate per voice channel	4800 bps	4800 bps	9600 bps	9600 bps
Mod Bandwidth	7.6 kHz	6.0 kHz	9.36 kHz	8.1 kHz
Channel bandwidth	12.5 kHz	6.25 kHz	12.5 kHz	12.5 kHz
Voice channels	2	1	1	1
Developer	ETSI	JARL	Yaesu	APCO

User information	DMR	D-Star	Fusion
Registration required	Yes	Yes	No
User ID transmitted	DMR ID*3	Callsign	Callsign
User ID displayed	DMR ID*4	Callsign	Callsign
Is the ID OK for FCC and other jurisdictions?	No	Yes	Yes

*1 C4FM (continuous envelope 4 frequency modulation) is a variant of 4FSK used by Yaesu System Fusion and P25 phase 1.

*2 GMSK (Gaussian Minimum Shift Keying) is a variant of FSK used for D-Star.

*3 most amateur radio licencing authorities require you to transmit your callsign. DMR does not do this automatically. You need to identify your transmission by saying your callsign, the same as you would on an FM repeater.

*4 the mobile, handheld, or hotspot will only display the other station's callsign if it is held in the contact list in your radio, or the Talker Alias includes the callsign. Otherwise only the ID number is displayed. There are around 206,000 registered DMR users. Some radios can store all of them. With other radios, you will have to be more selective. You can download selected data from https://www.radioid.net/. Or you may elect to only enter contact data from your local area, or for Hams you call often.

User repeater connections	DMR	D-Star	Fusion
Talk to local stations	Yes	Yes	Yes
Link to another repeater	No	Yes	No
Internet-linked systems	Talk groups	Reflectors	Wires-X rooms
Selection	Channel switch	UR field	Room name
Group Call	Yes	Yes	Yes
Private Call	Yes	Yes	No
Echo test	Yes	Yes	No

Ease of use	DMR	D-Star	Fusion
Memory selection	Channel switch	Dial or list	Channel switch
Analog mode	Programmed channel	Button	Button
Programming	Difficult*1	Medium	Easy
Using the radio	Easy	Medium	Easy
Radio models (mostly)	Many	Icom	Yaesu
Radio cost	Cheapest	Medium	Most expensive
Hot spots	Any MMDVM Pref Duplex	Any MMDVM	Any MMDVM or Wires-X
Send photos	No	Yes	No
SMS (text) messages	Yes	Yes	No
Send GPS/APRS data	Some models	Yes	No

*1 It's not that programming a DMR radio is difficult to accomplish. You just download a new code plug config file using a USB cable. The annoying thing that makes life a little difficult is that you may have to edit and download a new file any time you want to add a new talk group or repeater to the radio. Adding a new repeater can involve adding many additional lines to the file.

Troubleshooting

RADIO KEYPAD BUTTONS NOT WORKING

The keypad may be locked. See your radio manual for details on how to unlock it. The ID-52 and ID-51 have a keyboard lock. Press the **Menu** button to activate the menu function, then press and hold it again, to enable or disable the keyboard lock.

The IC-705 and IC-9700 have a lock button marked as a door key. It only locks the VFO dial. Press and hold it to lock or unlock the dial.

OUTPUT POWER IS LOW

Some manufacturers are prone to exaggerating the RF output power. More specifically the radio probably will transmit at the stated power in FM mode if the battery is at full charge, but the output power may be much lower if the battery is down a bit. **Don't worry about it!** Remember that transmitting at half power is only a 3 dB reduction in signal strength at the repeater. So, unless you are right on the fringe, running slightly less transmitted power will not make any difference.

Use the lowest power setting if you are transmitting to a hotspot. You don't need full power to get across the room. For repeaters, to conserve battery life, use the lowest power setting that works reliably. Using the radio at full power makes it run hot and dramatically increases battery consumption.

The power may look lower in the D-Star mode because the bandwidth of the transmitted signal is wider than an unmodulated carrier such as FM mode with no modulating audio.

CAN'T HEAR THE REPEATER

You have the D-Star channel programmed, and you can see transmissions on the RSSI (receive signal strength indicator), but you cannot hear anything. If the repeater is a 'multi-mode repeater.' It could be transmitting a P25, DMR, or Fusion signal. You will only be able to hear the transmission if it is a D-Star signal. Check you have the correct frequency and offset loaded. For FM repeaters check the tone squelch is correct.

RADIO RECEIVER PULSING

The radio receiver pulsing except when listening to an incoming call is an indication that it is receiving the repeater or hotspot but is not able to communicate with it. In the case of the repeater, it may be too far away, or you may not be running enough power. Possibly your transmit frequency is using an incorrect offset. In the case of a hotspot, pulsing usually means that the hotspot is off frequency, or that you have your Pi-Star frequency settings wrong.

The hotspot is repeatedly trying to get a response from your radio, hence the repeated pulsing. Check out 'Hotspot off frequency,' below.

I had this happen when I switched the hotspot from duplex to simplex, and then back to duplex. Although it reverted to having two frequency boxes, it preloaded them both with the hotspot's RX frequency. This meant that the radio could receive signals from the hotspot, but the hotspot could not receive signals from the radio.

DR BUTTON DOES NOT WORK

The digital repeater mode will not work if there are no repeaters listed in the DV repeater memory. You need at least your local repeater or hotspot in there.

HIGH BER INDICATED ON PI-STAR

High BER when you transmit shows up as a red indicator on the Local RF Activity and Gateway Activity screens on Pi-Star. It usually means that your hotspot is off frequency and an offset needs to be applied in Pi-Star, not in your radio! See 'Hotspot off frequency,' below. Received signals marked with red high BER or Loss on the Pi-Star 'Gateway Activity' indicates that the calling station has a problem. It could be that their hotspot is off frequency or indicate some Internet packet loss problem.

MY NEW HOTSPOT DOESN'T WORK

Don't panic! Most problems are easily solved. Sometimes just restarting the hotspot is enough. You are much more likely to have a problem with a simplex MMDVM hotspot than a duplex MMDVM hotspot. Other hotspots like OpenSpot, DVAP, etc. are less prone to problems and are better supported by the manufacturer. The most common problems and symptoms are,

- An insufficient power supply. Symptoms, include Windows 'bonging' regularly, hotspot rebooting, and no display or an incorrect display.
- The display is not configured correctly in Pi-Star. No display even when rebooting the hotspot.
- Incorrect Modem setting in Pi-Star. No transmission to the radio. No display.
- Network problems in Pi-Star.
- The hotspot is a long way off frequency. Seeing activity on the hotspot display but no audio from the radio, no display on the radio of calling stations, hotspot not responding when you transmit.
- The hotspot is a little off-frequency. Garbled incoming and/or outgoing speech. High BER or Loss % reported on the Pi-Star Gateway Activity screen.

Insufficient power supply

An insufficient power supply can be a problem, especially if you are using a Raspberry Pi 3 rather than a Raspberry Pi Zero W.

I have had no issues powering my hotspot from a USB3 port on my computer, but the recommendation from the Raspberry Pi Foundation is to use a 2.5 amp USB power supply or a USB battery bank.

Display configuration

The display might not work if there is insufficient power, or it is not configured correctly in Pi-Star.

For an OLED (small display), select OLED type 3 for the very small 0.96" screen or OLED type 6 for the more common 1.3" screen. Either works on my hotspot. The Port should be set to /dev/tty/AMA0 because the display is being driven directly by the modem board.

If you have a Nextion screen, select Nextion. The port is usually set to modem unless the display is connected to a TTL to USB adapter plugged into the Rpi. In that case select /dev/tty/USB0.

Incorrect Modem setting in Pi-Star.

This is indicated by no transmission to the radio and/or no hotspot display. Check the advertisement for the modem you have purchased and see if there is any indication of the modem setting that should be set in Configuration > General Configuration > Radio/Modem Type. My duplex modem is set to MMDVM_HS_Hat_Dual Hat (VR2VYE) for Pi (GPIO), as specified by the manufacturer.

Dual Hat means it is a duplex modem, VR2VYE is the person who wrote the firmware, it is for a Pi, and it is connected to the Pi using the GPIO header pins, not a USB interface.

I had trouble selecting the correct modem for my simplex hotspot, mostly because it was so far off frequency it was not working anyway.

There are several choices. Make sure that you choose an RPi version, rather than a DV-Mega, Nano or NPi version. Unless you do have one of those.

The STM32-DVM / MMDVM_HS - Raspberry Pi Hat (GPIO) modem seems to work with most generic Chinese simplex hotspots. Using some of the other choices displayed calls on the hotspot, but there was no signal received by the radio.

Network problems in Pi-Star

Check your Pi-Star dashboard page. The **Modes Enabled** section should have D-Star in green. The **Network Status** section should have D-Star Net in green. In the **D-Star**

Repeater section, RPT1 should show your callsign with a module extension of A, B, or C depending on the frequency of your hotspot. My dashboard says 'ZL3DW B' because I have a UHF (70cm) hotspot. RPT2 should show your callsign with a G for gateway extension. In the **D-Star Network** section, the APRS gateway is only required if you are using D-PRS. It is set in the General Configuration part of the Configuration tab. The last line shows the currently connected reflector, or 'Not Linked' if no reflector is currently connected.

There are other indicators on the '**Admin**' tab, in the **Service Status** area. MMDVDM host should be green, indicating that a Pi-Star hotspot is connected. ircDDBGateway should be green, indicating an Internet connection through your home LAN to the D-Star registry, and PiStar-Watchdog should be green indicating that the Pi-Star software is regularly communicating with the hotspot.

If any of those settings are wrong, go back through the Pi-Star setup.

HOTSPOT OFF FREQUENCY

This is the one that trips up most people, probably because it is the hardest to fix. It is very common for a new hotspot to be off-frequency. The frequency error should always be corrected in Pi-Star and never with the radio or CS offsets. There is no point in deliberately making your radio transmit off frequency to compensate for a frequency error in the hotspot. How do you know the problem is in the hotspot, not the radio? It is far more likely to be the cheapo hotspot than the expensive radio. The only way to check for sure would be to compare it with another radio.

TIP: Some hotspots are checked by the vendor and shipped with an offset number on a slip of paper in the box or stuck to the bottom of the modem. I was not lucky in that respect.

When I powered up my new simplex hotspot it would not respond to my handheld radio when I transmitted, and although I could see callers on the hotspot display and a LED and RSSI indication on the radio, I could not hear them. It wasn't until I watched a setup video on YouTube, I remembered the hotspot frequency offsets. I am fortunate to own a spectrum analyser and a good frequency counter, so I was able to quickly find out that my hotspot was a whopping 4.4 kHz off frequency. The online video was talking about a frequency error of only 472 Hz.

First check

Before you attempt to change the hotspot offset, make very sure that the radio frequencies match the hotspot frequencies.

Select the Pi-Star Configuration tab. If you are using a simplex hotspot, The Controller Mode in the Control Software should be set to Simplex Node. The Controller Software should be set to MMDVM Host. The radio channel should be set for simplex, and it should match the hotspot frequency.

If you are using a duplex hotspot, The Controller Mode in the Control Software should be set to **Duplex Repeater**. The Controller Software should be set to **MMDVM Host**. The frequencies in the General Configuration should be different by 5 MHz on 70cm or 600 kHz on 2m. Check the math! Check if it is a positive or negative offset.

The channel in your transceiver should be set for duplex with the correct offset and should match the hotspot frequencies.

Radio Frequency RX: is the hotspot's receive frequency. It should be set to the transceiver's transmit frequency. I use the normal repeater offset for the band. 5 MHz for the 70cm band or 600 kHz for the 2m band.

Radio Frequency TX: is the hotspot's transmit frequency. It should be set to the transceiver's receive frequency. This is the frequency programmed into the DR memory and displayed on the radio while it is receiving.

If you still have a problem, you can set the hotspot frequency offset.

Setting the frequency offset

You can be pretty sure that any offset applied to the hotspot's transmit frequency will also be correct for its receive frequency. They are generated from the same oscillator. I always set both offsets the same.

There are two ways you can do this. Either adjust the receiver offset for the best BER (bit error rate) when you transmit to the hotspot. Or adjust the transmit offset while the hotspot is transmitting, by measuring the frequency it is radiating with a spectrum analyser, frequency counter, or SDR receiver. If the hotspot is seeing the transmission from your radio and the dashboard is showing 'Local RF Activity' use option 1. If the hotspot is not seeing the transmission from your radio and the dashboard is not showing 'Local RF Activity,' but is showing 'Gateway Activity,' use option 2. The option 2 method is faster and more accurate, but it requires test equipment. Option 1 is a perfectly acceptable method.

Option 1: adjust the offset for the best BER

This method only works if the hotspot is seeing your transmission and it is showing up on the 'Local RF Activity' section of the Pi-Star dashboard page.

TIP: If it is not and you do not have a frequency counter, spectrum analyser, or SDR receiver capable of receiving the 70cm band you can try transmitting at a range of offsets until you can get into the hotspot. See option 3.

Open Pi-Star and look at the 'Dashboard' page. Specifically, the BER (bit error rate) indication in the 'Local RF Activity' section.

1. Set your radio to your hotspot frequency and select Local CQ in the TO Select box so that you won't annoy anyone.

2. Key up your radio with the PTT and hold it on transmit for four or five seconds. You should see your callsign pop up in the Local RF Activity area and at the top of the Gateway Activity region. When you release the PTT, you should see an indication on the BER meter (hopefully 0%) and the SRC box should be green with 'RF' in it. If it is green the frequency offset is good. Job done. If it is red, the frequency offset needs to be adjusted.

   To adjust the offset in Pi-Star, select **Configuration > Expert > MMDVM Host**. Scroll down to the **Modem** section and find **RXoffset** and **TXoffset**.
   (Not RXDCoffset and TXDCoffset).

3. Set both **RXoffset** and **TXoffset** to +500 Hz. Click **Apply Changes**.

4. After the hotspot reboots, transmit again and note if the BER has improved.

   a. If it got worse, the frequency may be high. Set both **RXoffset and TXoffset** to -500 Hz and try again.

   b. If it got better, increase the offset and try again. Keep changing both offsets, using smaller and smaller changes, until the BER is less than 0.5%.

Option 2: adjust the hotspot using its transmitter frequency

Using a frequency counter or spectrum analyser is more accurate than using the BER method, but it does require you to have the test equipment.

Remember to use a 20 dB coaxial RF attenuator in the cable if you are connecting your hotspot to a frequency counter or spectrum analyser.

1. Select a busy reflector like REF001 C, so that there is plenty of activity on the channel.

2. Take note of the hotspot's transmitter (TX) frequency or the simplex frequency. It is in the Radio Info box on the Pi-Star dashboard.

3. In the Pi-Star software select **Configuration > Expert > MMDVM Host**. Scroll down to the **Modem** section and find **RXoffset** and **TXoffset**.
   (Not RXDCoffset and TXDCoffset).

4. Monitor the hotspot and observe the frequency when it transmits. Subtract the wanted frequency as indicated on the Pi-Star dashboard from the observed hotspot transmitter frequency.

5. Adjust both the **RXoffset** and **TXoffset** by the amount of offset needed in Hertz. Click **Apply Changes**. After the hotspot reboots, observe the frequency when the hotspot transmits. Make any minor adjustments. You should be able to get to within about 100 Hz. The adjustment process is not fine enough to get the frequency exactly right.

6. You should make a final check that the BER is good when you transmit towards the hotspot, using the instructions in option 1: The BER indication should be less than 0.5% and green.

	Modem
Port	/dev/ttyAMA0
TXInvert	1
RXInvert	0
PTTInvert	0
TXDelay	100
RXOffset	0 Adjust frequency
TXOffset	0 Adjust frequency
DMRDelay	0
RXLevel	50
TXLevel	50
RXDCOffset	0 Not this one
TXDCOffset	0 Not this one

Using an SDR receiver as a spectrum analyser

If you don't have a frequency counter or spectrum analyser, you can use an SDR receiver to do the test. I do not recommend a direct connection. Attach an antenna to the SDR. It should be sensitive enough to see the signal from the hotspot. The frequency accuracy of your SDR should be good enough to determine if the modem is transmitting a long way off frequency.

Option 3, the 'hit and miss,' method.

If option one does not work and you do not have any test gear, you can use the 'trial and error' method. Set a 500 Hz offset and see if you can hear traffic on the hotspot or if it displays your callsign when you transmit towards it. Keep stepping up (or down) in 500 Hertz steps until you hopefully have some success. I think it is unlikely the hotspot will be more than 5000 Hz high or 5000 Hz low. It will be slow going as you have to Apply Changes each time you change the offset, but eventually, it should get the hotspot close enough for the callsign to display and indicate a green RF in the Src box.

Local RF Activity

Time (NZST)	Mode	Callsign	Target	Src	Dur(s)	BER	RSSI
10:00:02 Sep 21st	D-Star	ZL3DW/9700 (GPS)	CQCQCQ	RF	64.0	0.0%	S9+46dB (-47 dBm)

CS PROGRAMMING SOFTWARE WON'T WORK

If you have an old version of the CS software, it may refuse to communicate with the radio. I had this happen and it was confusing. Download the latest version of the CS software for your radio model, from the Icom website.

Internet links

The Internet is quite large. I can't mention the thousands of links to all the possible D-Star references. Here are a few that I found useful.

D-Star Info has a list of REF reflectors http://www.dstarinfo.com/reflectors.aspx It also has repeater lists, downloads for D-Star DV repeater lists, Nets, and loads of other information.

The 'DARA Thursday Night Group' website has links to 86 D-Star REF reflectors https://sites.google.com/site/darathursdaynite/d-star/d-star-dplus-reflectors

There is a list of XRF reflectors at http://xrefl.net/.

A good list of active XLX reflectors http://www.xlx750.nz/index.php?show=reflectors

D-Star repeaters, gateways, last heard https://www.dstarusers.org/

Hotspot software and a lot of other reference information https://www.pistar.uk/

Last heard monitor (D-Star activity) https://nj6n.com/dplusmon/

Last heard monitor (D-Star activity) www.ircddb.net/live.htm

Last heard monitor (D-Star activity) https://www.dstarusers.org/lastheard.php

D-Star Australia resource https://www.dstar.org.au/about/resources/

XLX014 / DCS014 dashboard http://dcs014.xreflector.net/index.php

REF023 dashboard http://ref023.dstargateway.org/

XLX299 dashboard http://www.xlx299.nz/

XFR750 dashboard http://www.xlx750.nz/

REF001 dashboard http://ref001.dstargateway.org/

D-STAR QUICK-START GUIDE by Rob Locher W7GH
http://www.roblocher.com/whitepapers/dstar.html

Amateur Radio Notes by Toshen KE0FHS
https://amateurradionotes.com/d-star.htm

QuadNet https://www.openquad.net/groupsv6.php

D-Star users.org https://www.dstarusers.org/

Videos

Many people have created excellent videos relating to D-Star. Sometimes seeing someone perform a task is easier than reading about it. Here are a few that I liked. If you like a video on this list, please hit the 'Like' button and subscribe to the channel. It encourages the creators and helps pay for more great videos.

Discover D-Star (Icom)
https://www.youtube.com/watch?v=ZMMt55Dtp5Q

Icom D-Star registration (Icom USA)
https://www.youtube.com/watch?v=cPp8DHB9arQ

D-Star and DMR for beginners – radios, hotspots and more (K6UDA)
https://www.youtube.com/watch?v=QfkHwgourDg&t=952s

Introduction to D-Star (Gadget Talk)
https://www.youtube.com/watch?v=yN4L4YTtLtE

Build your own DMR/DStar/Fusion hotspot for CHEAP (AD6DM Dennis)
https://www.youtube.com/watch?v=LspgnvDPJvc

Quick, let's get on D-Star Terminal Mode (Steve KM9G)
https://www.youtube.com/watch?v=8u9ZiwQfNOI

The simple guide to using D-Star radio (The Ham Radio Junkie)
https://www.youtube.com/watch?v=GBFebJHk3wc

How does D-Star work (Jonathon M0JSX)
https://www.youtube.com/watch?v=-4OP0XHeZOk

The difference between DV mode and DR mode (K9WLW Radio)
https://www.youtube.com/watch?v=FFDcOOO6HBY

Portable AMBE Server – no radio D-Star (VK4NGA)
https://www.youtube.com/watch?v=TfQEt3-BtDk

D-Star Basic Radio Programming (RT Systems)
https://www.youtube.com/watch?v=ZAdizwHKpDI

Turn your pi-star hotspot into a mini DStar Repeater! (K9WLW Radio)
https://www.youtube.com/watch?v=CkY9PrroE58

Linking Pi-Star Hotspots with D-Star (David Cappello)
https://www.youtube.com/watch?v=g28xAKt3Vkg

Icom ID-52 Handheld, RS-MS1A App (Ham Radio Concepts)
https://www.youtube.com/watch?v=_RylMv4jPV8

Glossary

59	Standard (default) signal report for amateur radio voice conversations. A report of '59' means excellent readability and strength.
73	Morse code abbreviation 'best wishes, see you later.' It is used when you have finished transmitting at the end of the conversation.
.dll	Dynamic Link Library. A reusable software block that can be called from other programs.
2m, 70cm	Two metre (144 MHz) and 70cm (430 MHz) amateur radio bands
A/D	Analog to digital
ADC	Analog to digital converter or analog to digital conversion
AF	Audio frequency - nominally 20 to 20,000 Hz.
Algorithm	A process, or set of rules, to be followed in calculations or other problem-solving operations, especially by a computer. In DSP it is a mathematical formula, code block, or process that acts on the data signal stream to perform a particular function, for example, a noise filter.
AMBE+2	Advanced Multi-Band Excitation version 2. The AMBE+2 Vocoder uses a propriety chip made by DVSI (Digital Voice Systems Incorporated) to convert speech into a coded digital signal or the received digital signal back to speech. Can transmit intelligible speech with data rates as low as 2 kbs. Used for D-Star, DMR, YSF, NXDN, and 'phase 2' P25.
APCO	Association of Public-Safety Communications Officials
APRS	Automatic packet reporting system – used to send and display location information from a GPS receiver. APRS beacons transmitted over D-Star are displayed on the APRS.fi website.
BER	Bit error rate – a quality measurement for any digital transmission system. It measures the number of bits that were received incorrectly compared to the overall bit rate.
Bit	Binary value 0 or 1.
bps	Bits per second (data speed)
BW	Bandwidth. The range between two frequencies. For example, an audio passband from 200 Hz to 2800 Hz has a 2.6 kHz bandwidth.
C4FM	Continuous 4-state Frequency Modulation (used for P25 Phase 1 and Fusion)
Carrier	Usually refers to the transmission of an unmodulated RF signal. It is called a carrier because the modulation process modifies the un-modulated RF signal to carry the modulation information. A carrier

	signal can be amplitude, frequency, and/or phase modulated. Then it is referred to as a 'modulated carrier.' An oscillator signal is not a carrier unless it is transmitted.
CODEC	Coder/decoder - a device or software used for encoding and decoding a digital data stream.
Colour Code	The colour code is used to identify the output of a specific DMR repeater, much like the CTCSS tone on an FM repeater. It is not relevant to D-Star operation.
CPU	Central processing unit. The ARM (advanced RISC machine) processor in the Raspberry Pi, or the microprocessor in your PC. [RISC is reduced instruction set computing, an acronym inside an acronym.]
Cross-connect	A link between different technologies or networks. Such as a D-Star reflector linked to a System Fusion 'Room' or a DMR 'Talk Group'
CTCSS	Continuous Tone Coded Squelch System, used for access control to most analog FM repeaters and FM handheld or mobile radios.
D/A	Digital to analog.
DAC	Digital to analog converter or digital to analog conversion
Dashboard (Pi-Star)	The Pi-Star dashboard is an HTML website hosted on the Pi-Star hotspot and accessed via any web browser on the same WiFi or local network. It displays traffic being passed through the hotspot and the hotspot's configuration settings.
Dashboard (repeater)	A repeater dashboard is an HTML website that displays the status of a repeater or Hotspot, including what frequencies it is on, its location, who owns it, who is using it, and what talk groups it is linked to. Most repeater dashboards are available to the amateur radio community, or selected people, over the Internet. You can choose whether to make your Hotspot public or private.
data	A stream of binary digital bits carrying information
dB, dBm, dBc, dBV	The Decibel (dB) is a way of representing numbers using a logarithmic scale. Decibels are used to describe a ratio, the difference between two levels or numbers. They are often referenced to a fixed value such as a Volt (dBV), a milliwatt (dBm), or the carrier level (dBc). Decibels are also used to represent logarithmic units of gain or loss. An amplifier might have 3 dB of gain. An attenuator might have a loss of 10 dB.
DC	Direct Current. The battery or power supply for your radio, charger, or Hotspot will be a DC power supply.
Direct Mode	is another name for direct radio-to-radio contact on a simplex frequency.

DMR	DMR stands for Digital Mobile Radio. A standard for sending voice traffic over a digital radio link to another radio, through a repeater, or to a station connected to a connected Talk Group.
DMR+	DMR+ is a worldwide DMR network. It was the first to interconnect ETSI standard Tier II repeaters. It is aligned with the DMR-MARC network so you can access the DMR-MARC Talk Groups. The DMR+ network specialises in interconnections with other technologies such as D-Star, AllStar, and C4M (P25 and YSF).
DMR-MARC	DMR-MARC is a network of DMR repeaters established by the Motorola Amateur Radio Club. The members of MARC were instrumental in getting DMR established for amateur radio. They set up the first amateur radio DMR networks and repeaters. There are around 500 DMR-MARC repeaters in 83 countries with over 144,000 registered users.
D-Star	Digital Smart Technologies for Amateur Radio. D-Star is the (mostly) Icom digital voice system. Unlike DMR it was developed specifically for amateur radio.
DVSI	The AMBE+2 Vocoder uses a propriety chip designed and made by DVSI (Digital Voice Systems Incorporated) to convert speech into a coded digital signal or the received digital signal back to speech.
Duplex	A radio or Hotspot that can receive and transmit at the same time. Usually on different frequencies. A standard repeater is a duplex system.
DX	Long-distance, or rare, or wanted by you, amateur radio station. The abbreviation comes from the Morse telegraphy code for 'distant exchange.'
Echo	A D-Star reflector, usually the E extension, that repeats back a test call that you make.
ESSID	Extended service set identifier – a 2-digit extension to your DMR ID number to identify a second or subsequent hotspot on the same network.
ETSI	The European Telecommunications Standards Institute developed the DMR platform.
FM	Frequency modulation. The "good ol'" analog repeater system.
FSK	Frequency Shift Keying. DMR transmitters use 4-state frequency shift keying modulation. Each frequency shift carries two bits of the input data stream
FTDI	USB to 3.3V TTL level converter designed by Future Technology Devices International Ltd.
GMSK	Gaussian Minimum Shift Keying - spectrum efficient frequency shift keying mode used for D-Star

GPS	Global Positioning System. A network of satellites used for navigation, geolocation, and very accurate time signals.
Hex	Hexadecimal – a base 16 number system used as a convenient way to represent binary numbers. For example, 1001 1000 in binary is equal to 98h or 152 in decimal.
Hotspot (D-Star)	A DMR Hotspot is a small Internet connected box that can connect to D-Star reflectors. You transmit from your D-Star handheld to the Hotspot, and it passes the data through to the Internet. The information that is returned is transmitted by the Hotspot back to your radio. 77% of all D-Star hotspots are MMDVM. The rest are mostly OpenSpot or DVMega hotspots.
Hotspot (WiFi)	Many cell phones can be configured to act as a WiFi hotspot, allowing WiFi devices to get access to the Internet via your phone and mobile data plan. You could connect a WiFi-enabled D-Star hotspot to a WiFi hotspot on your phone and connect your D-Star radio to worldwide reflectors via your phone.
Hz	Hertz is a unit of frequency. 1 Hz = 1 cycle per second.
IMBE	Improved Multi-Band Excitation. Vocoder used for P25 phase 1 digital voice
JARL	Japan Amateur Radio League. The official Japanese amateur radio organisation. Developed D-Star in association with Icom.
kbits	Thousands of bits per second. 1 kbit – 1000 bps.
kHz	Kilohertz is a unit of frequency. 1 kHz = 1 thousand cycles per second.
LAN	Local Area Network. The Ethernet and WIFI-connected devices connected to an ADSL or fibre router at your house are a LAN.
LED	Light Emitting Diode
LoTW	Logbook of the World. An ARRL QSO logging database that is used worldwide.
Mbits	Millions of bits per second (data rate)
MHz	Megahertz – unit of frequency = 1 million cycles per second.
MIC	Microphone
MMDVM	Multi-mode digital voice modem - usually supports DMR, D-Star, Fusion, P25, and NXDN. 77% of all DMR Hotspots are MMDVM Hotspots. The rest are mostly OpenSpot or DVMega.
Module	A suffix that follows the callsign that tells the repeater and D-Star network what type of item is connecting or being connected to a reflector or gateway. Reflectors use them. Gateways use them and your radio uses them. A=1.2GHz band, B=70cm band, C= 2m band, G= Gateway. The rest are used mostly on reflectors to indicate different subgroups on the same reflector.

MOTOTRBO	MOTOTRBO is a Motorola trademark used to describe their range of DMR products
Network (DStar)	The D-Star network is a collection of interconnected 'gateway' repeaters and reflectors.
Onboard	A feature or data list that is contained within the radio.
Parrot	A colourful class of birds known for their ability to mimic speech and other sounds. In DMR it is a system that repeats back a test call made from your DMR. Known as Echo on D-Star.
PC	Personal Computer. For the examples throughout this book, it means a computer running Windows 10.
PSK	Pre-shared key – a password or security code known to your router and a connected router. For example, your BM password.
PTT	Press to talk - the transmit button on a microphone – pressing the PTT makes the radio transmit.
QSO	Q code – an amateur radio conversation or "contact."
QSY	Q code – a request or decision to change to another frequency.
POTA	Parks on the Air
RF	Radio Frequency
RS232	A computer interface used for serial data communications.
RPi	Raspberry Pi single-board computer
RX	Abbreviation for receive or receiver
Simplex	Simplex means to receive and transmit on the same frequency. In most cases, a radio operating on a simplex frequency cannot transmit and receive simultaneously. Simplex can be used if you wish to communicate directly with another D-Star radio (without a repeater or the Internet).
Simplex Repeater or Hotspot	A simplex repeater uses the same frequency for receiving and transmitting. It passes data (digital voice) from an Internet connection to the hotspot or repeater transmitter so that you can receive it on your radio. When you transmit, the simplex repeater or hotspot passes the signal received from your radio to a talk group over the Internet connection.
SOTA	Summits on the Air
Split	The practice of transmitting on a different frequency to the one that you are receiving on. Repeaters use a 'repeater split' between the repeater input frequency and the repeater output frequency.
Squelch	Squelch mutes the audio to the speaker when you are in FM mode and not receiving a wanted signal. When a signal (with the correct CTCSS tone, if enabled) is received, the squelch 'opens' and you can hear the station.
SSID	Service set identifier – in Pi-Star it is your WiFi network name

Tail	Furry attachment at the back of a dog or cat. Also, the length of time an FM repeater stays transmitting after the input signal has been lost. It can also mean a short flexible length of coaxial cable at the antenna or shack end of your main feeder cable.
Talk Group	A Talk Group is the DMR equivalent of a D-Star 'reflector' or a Fusion 'Room.' It is a collection of linked repeaters that are configured so that users with a common interest or from a common location can talk to each other. For example, there are Worldwide, North American, State, and County Talk Groups. Also, Spanish Language and Old Timers groups.
TDMA	Time domain multiple access. A technique for interleaving the data from two or more voice or data channels onto a single data stream. DMR uses TDMA to combine two voice (and data) Time Slots onto a data stream, which is used to modulate the radio using 4FSK.
TX	Abbreviation for Transmitting or Transmitter.
UHF	Ultra-High Frequency (300 MHz - 3000 MHz).
USB	Universal serial bus – serial data communications between a computer and other devices. USB 2.0 is fast. USB 3.0 is very fast.
VFO	Variable Frequency Oscillator. This applies to radios that can be tuned in frequency steps rather than stepping through previously saved memory channels.
VHF	Very High Frequency (30 MHz -300 MHz)
Vocoder	A Vocoder is a category of voice codec that analyses and synthesizes the human voice signal for audio data compression, multiplexing, voice encryption or voice transformation. The vocoder was invented in 1938 by Homer Dudley at Bell Labs as a means of synthesizing human speech. [Wikipedia]
W	Watts – unit of power (electrical or RF).
YSF	Yaesu System Fusion

Table of drawings and images

Index

The Author

Well, if you have managed to get this far you deserve a cup of coffee and a chocolate biscuit. It is not easy to digest large chunks of technical information. It is probably better to dip into the book as a technical reference. Anyway, I hope you enjoyed it and that it has made learning about D-Star a little easier.

I live in Christchurch, New Zealand. I am married to Carol who is very understanding and tolerant of my obsession with amateur radio. She describes my efforts as "Andrew playing around with radios." We have two children and two cats. James has graduated from Canterbury University with a degree in Commerce and is working for a large food wholesaler. Alex is a doctor working in the Christchurch Hospitals.

I am a keen amateur radio operator who enjoys radio contesting, chasing DX, digital modes, and satellite operating. But I am rubbish at sending and receiving Morse code. I write extensively about many aspects of the amateur radio hobby.

Thanks for reading my book!

73 de Andrew ZL3DW.